Lecture Notes
in Business Information Processing 193

Stanisław Wrycza (Ed.)

Information Systems: Education, Applications, Research

7th SIGSAND/PLAIS EuroSymposium 2014
Gdańsk, Poland, September 25, 2014
Proceedings

 Springer

Volume Editor

Stanisław Wrycza
University of Gdańsk
Department of Business Informatics
Sopot, Poland
E-mail: swrycza@univ.gda.pl

ISSN 1865-1348　　　　　　　　　e-ISSN 1865-1356
ISBN 978-3-319-11372-2　　　　　　e-ISBN 978-3-319-11373-9
DOI 10.1007/978-3-319-11373-9
Springer Cham Heidelberg New York Dordrecht London

Library of Congress Control Number: 2014948361

Typesetting: Camera-ready by author, data conversion by Scientific Publishing Services, Chennai, India

Printed on acid-free paper

Springer is part of Springer Science+Business Media (www.springer.com)

Preface

Systems analysis and design (SAND) has been the central field of research and education in the area of management information systems (MIS) or, as it is called more frequently in Europe, business informatics, almost from its origins. SAND continuously attracts the attention of both academia and business. The rapid progress of information and communications technology naturally generates the requirements for new generations of SAND methods, techniques, and tools. Therefore, international thematic conferences and symposia have become widely accepted forums for exchanging of concepts, solutions, and experiences in SAND. In particular, the Association for Information Systems (AIS) is undertaking the initiative toward the international development of SAND.

The objective of the EuroSymposium on Systems Analysis and Design is to promote and develop high-quality research on all issues related to SAND. It provides a forum for SAND researchers and practitioners in Europe and beyond to interact, collaborate, and develop their field. The EuroSymposia were initiated by Prof. Keng Siau as the SIGSAND - Europe Initiative. Previous EuroSymposia were held at:

- University of Galway, Ireland – 2006
- University of Marburg, Germany – 2008
- University of Gdansk, Poland – 2007, 2011, 2012, 2013

The accepted submissions of EuroSymposia were published as:

- EuroSymposium 2007: A. Bajaj, S. Wrycza (eds.), Systems Analysis and Design for Advanced Modeling Methods: Best Practises, Information Science Reference, IGI Global, Hershey, New York, 2009
- EuroSymposium 2011: S. Wrycza (ed.) 2011, Research in Systems Analysis and Design: Models and Methods, series: LNBIP 93, Springer, Berlin 2011
- Joint Working Conferences EMMSAD/EuroSymposium 2012 held at CAiSE'12: I. Bider, T. Halpin, J. Krogstie, S. Nurcan, E. Proper, R. Schmidt, P. Soffer, S. Wrycza (eds.) 2012, Enterprise, Business-Process and Information Systems Modeling, series: LNBIP 113, Springer, Berlin 2012
- 6th SIGSAND/PLAIS EuroSymposium 2013: Stanisław Wrycza (ed.), Information Systems: Development, Learning, Security, Series: Lecture Notes in Business Information Processing 161, Springer, Berlin 2013

There were three organizers of the 7th EuroSymposium on Systems Analysis and Design, as follows:

- SIGSAND – AIS Special Interest Group on Systems Analysis and Design
- PLAIS – the Polish Chapter of AIS
- The Department of Business Informatics of University of Gdansk, Poland

SIGSAND is one of the most active AIS SIGs with a substantial record of contributions to AIS. It provides services such as the annual North American and European SAND Symposia, research and teaching tracks at major IS conferences, a listserv, and special issues in journals.

The Polish Chapter of the Association for Information Systems (PLAIS) was established in 2006 as the joint initiative of Prof. Claudia Loebbecke, former President of AIS, and Prof. Stanisław Wrycza, University of Gdansk, Poland. PLAIS co-organizes international and local IS conferences.

The Department of Business Informatics of the University of Gdansk is conducting intensive teaching and research activities. Some of its academic manuals are bestsellers in Poland. The Department is also active internationally, organizing conferencs including the 10th European Conference on Information Systems (ECIS 2002), the 7th International Conference on Perspectives in Business Informatics Research (BIR 2008), the 8th International Conference on European Distance and E-learning Network (EDEN 2009), and the 24th Conference on Advance Information Systems Engineering (CAiSE 2012). The department is the partner of the European Research Center for Information Systems consortium.

EuroSymposium 2014 had an acceptance rate of 40%, with submissions divided into the following two groups:

- Information Systems Education
- Information Systems Applications
- Information Systems Research

The accepted papers reflect the current trends in the field of SAND.

During EuroSymposium 2014, the following keynote speeches were given:

Claudia Loebbecke, Director of Department of Media and Technology Management, University of Cologne, Germany
"Big Data Analytics Impacting Teaching and Research"

Ricardo Ferreira, (European Union, DG Education and Culture, Unit A3 - Skills and Qualification)
"Opening up Education for Innovative Teaching and Learning in Europe"

I would like to express my thanks to all authors, reviewers, advisory board, international Program Committee and Organizing Committee members for their support, efforts, and time. They made possible another successful Systems Analysis and Design EuroSymposium.

July 2014   Stanisław Wrycza

Organization

General Chair

Stanisław Wrycza   University of Gdansk, Poland

Advisory Board

Wil van der Aalst   Eindhoven University of Technology,
      The Netherlands
Joerg Becker   University of Münster, Germany
Juhani Iivari   University of Oulu, Finland
Dimitris Karagiannis   University of Vienna, Austria
Julie Kendall   Rutgers University, USA
Claudia Loebbecke   University of Cologne, Germany
Keng Siau   Missouri University of Science and Technology,
      USA

Program Committee

Stanisław Wrycza   University of Gdańsk, Poland - Chair
Eduard Babkin   Higher School of Economics, Moscow, Russia
Akhilesh Bajaj   University of Tulsa, USA
Palash Bera   Saint Louis University, USA
Petr Doucek   University of Economics, Prague,
      Czech Republic
Rolf Granow   Lübeck University of Applied Sciences,
      Germany
Vijay Khatri   Indiana University Bloomington, USA
Marite Kirikova   Riga Technical University, Latvia
Andrzej Kobylinski   Warsaw School of Economics, Poland
Jolanta Kowal   University of Wroclaw, Poland
Tim Majchrzak   University of Cologne, Germany
Yannis Manolopoulos   University of Thessaloniki, Greece
Jinsoo Park   Seoul National University, Korea
Nava Pliskin   Ben-Gurion University of the Negev, Israel
Mijalce Santa   Ss. Cyril and Methodius University, Macedonia
Thomas Schuster   Forschungszentrum Informatik, Karlsruhe,
      Germany
Piotr Soja   Cracow University of Economics, Poland
Angelos Stefanidis   University of Glamorgan, UK

Reima Suomi	University of Turku, Finland
Jan Vanthienen	Catholic University of Leuven, Belgium
H. Roland Weistroffer	Virginia Commonwealth University, USA
Iryna Zolotaryova	Kharkiv National University of Economics, Ukraine
Joze Zupancic	University of Maribor, Slovenia

Organizing Committee

Chair:Stanisław Wrycza
Secretary: Anna Węsierska
Members: Dorota Buchnowska, Bartłomiej Gawin, Bartosz Marcinkowski, Jacek Maślankowski, Michał Kuciapski
Webmaster: Łukasz Malon

SIGSAND 2014 Topics

- Agile Methods
- Business Process Modeling
- Conceptual Modeling
- Curriculum Design and Implementation Issues
- Design of Mobile Applications
- Design Theory
- Empirical Studies in SAND
- Ethical, Human and Organizational Aspects of IS Development
- Human–Computer Interaction and Prototyping
- Information Systems Development: Methods and Techniques
- IT Design for Disaster Recovery
- Model-Driven Architecture
- Modeling Languages
- Ontological Foundations of Systems Analysis and Design
- Open Source Software (OSS) Solutions
- Rapid Systems Development
- Requirements Engineering
- SAND in ERP and CRM Systems
- SOA and Cloud Computing
- Systems Analysis and Design in Nonprofit Organizations
- Teaching Systems Analysis and Design
- UML/SysML/BPMN
- User Participation in SAND
- Web Design
- Workflow Management

Table of Contents

Table of Contents

IS Education Research: Review of Methods of Surveying the IS Curriculum to Support Future Development of IS Courses

Angelos Stefanidis[1] and Guy Fitzgerald[2]

[1] Faculty of Computing, Engineering and Science,
University of South Wales, Treforest, CF37 1DL, UK
angelos.stefanidis@southwales.ac.uk
[2] Schoold of Business and Economics, Loughborough University,
Loughborough, Leicestershire, LE11 3TU, UK
g.fitzgerald@lboro.ac.uk

Abstract. The work presented in this paper focuses on the relevant research approaches capable of supporting empirical survey studies regarding the analysis of IS curricula. The paper briefly considers the evolution of curriculum analysis as part of the wider IS Education context. Curriculum surveys have been used repeatedly for the purpose of exploring in detail the provision of IS knowledge and skills, and thus supporting academics in their efforts to update curricula in ways that reflect the needs of industry. In order to support research in this important area this paper reviews the principal data gathering techniques used for the benefit of future researchers engaging in similar work. Finally, it highlights the important factors affecting IS curriculum surveys.

Keywords: IS Education, IS Curriculum, Curriculum Surveys, Curriculum Mapping.

1 Introduction

For most of the past fifteen years academia and industry have been reporting decreasing numbers of IS undergraduate students and IS graduate professionals respectively. The largely inconclusive attempts made by academic researchers hoping to identify the exact reasons for this phenomenon, have highlighted issues ranging from the field's immaturity to a lack of meaningful cooperation between academia and industry. Among the possible explanations offered, a number of researchers have argued that ageing IS undergraduate curricula are partly to blame for the demise of the discipline [1, 2].

Much of the limited research in IS education examining IS curricula has taken the approach of investigating the perspective of one stakeholder (employers, academics, students, professional bodies, government) in order to frame a particular problem and suggest improvements [3-6]. The use of surveys as a research tool investigating IS curricula and job skills has been well established over the last twenty years [7, 8].

S. Wrycza (Ed.): SIGSAND/PLAIS EuroSymposium 2014, LNBIP 193, pp. 1–11, 2014.

Although the majority of the surveys undertaken during this time have focused on investigating the position of a specific stakeholder, the driving force behind most surveys has invariably been the potential alignment of the differing stakeholder positions [9].

As one of the prominent stakeholders, IS academics are predominantly concerned with IS research and teaching. The former shapes the direction of the field by considering both its applied and theoretical dimensions, often giving rise to arguments about balancing focus between them. The latter considers the curriculum development, teaching methods, and skills in relation to their relevance to industry. Interest in this issue of IS curriculum modernity (teaching) still remains strong because of the reduction in numbers of IS university applicants. Research in the area of IS student enrolments echoed the earlier expressed need for updating IS curricula [10-12]. At the same time, the body of work promoting the importance of possessing the right skills to succeed in the changing IS job market, continues to grow [13, 14]. In a recent article Benamati, Ozdemir [15] carried out a survey to measure the level of IS curricula alignment with industry needs, and examined the amount of change in the curriculum over a period of three years by reviewing a sample of IS courses in US universities. The authors concluded that while progress was being made, more work to address the problems of low student recruitment and insufficient graduate skills was necessary.

This paper considers the IS education research approaches relevant to IS curriculum surveys. Initially, the discussion focuses on past curriculum surveys by examining the merits of different data collection methods used. In essence, the work presented here considers the background research relating to IS curriculum surveys and the approaches which characterise it. The aim of the paper is to support IS academics, and management, in their efforts to analyse existing IS courses before they introduce changes intended to better reflect the prevailing needs of industry.

2   IS Education Research

A historical overview of the development of IS showed that during its early days, IS was characterised by a lack of clear research methodologies and a limited range of research themes which mostly focused on the identity of the fledgling discipline [16]. Research into the curriculum and wider issues affecting the educational aspects of IS was extremely narrow.

Such a state of affairs was not surprising given the absence of core research activity and the limited numbers of researchers who could identify directly with IS. Research in the field began to gather momentum as a result of early interest in what was perceived to be an applied social science with an implicit interest in technology and its use [17]. The early IS researchers were, in effect, 'stepping out' of their IS reference disciplines, such as computing, management and sociology, to do work in a field that lacked a clear identity but offered opportunities for innovative applied work. In doing so, the early IS research outputs were characterised by the approaches and methods of the reference domains that were helping IS form as an independent field.

It took over 20 years before [18] offered the first concrete declaration of an IS research classification scheme that provided a description of IS and the research areas that would develop it further [17]. IS education became part of this classification but its presence was much less prominent than other aspects of the field. Apart from providing context and direction, Barki's classification was a clear sign that IS research was becoming more focused. The continued growth in the pluralism of research approaches in the field gave rise to a mixture of theories, philosophical perspectives, research paradigms, methods and topics.

As the number of high quality research outlets began to grow, IS research aligned to different paradigms became prominent. Positivist IS research saw propositions and hypotheses being tested, along with models based on quantifiable data that stood the test of careful scrutiny. At the same time, interpretive research focused on non-deterministic perspectives about cultural phenomena which became an increasingly important part of the field [19, 20]. With the development of new research directions the methods used to conduct research grew. Case studies, surveys, experiments, action research, grounded theory and others were used to observe, study, analyse and influence the phenomena preoccupying the field of IS [21, 22]. Many theories and theoretical constructs supported the tens of research topics which emerged as the level of research maturity in the field continued to grow [23]. In the backdrop of these significant developments, research approaches specifically in IS education remained fairly unchanged, arguably due to the limited attention given to this area of IS.

3 Conducting IS Curriculum Surveys

Early attempts to analyse the IS curriculum begun with the undertaking of course surveys whose primary purpose was the cataloguing of modules and the skills they promoted. As such, the use of unobtrusive survey methods to gather and analyse information about the evolution of courses has a relatively long history in the IS field, dating back to the early days of its birth. The reasons that continue to fuel, even today, the need for measuring, mapping and classifying the IS curriculum stem from the somewhat tempestuous history of the field which, since its early days, has had to justify its existence and establish its legitimacy. The various identify crises of the field have caused many researchers to scrutinise the positioning of IS within the wider field of computing by examining its research and curriculum [16, 17]. Many of the discussions have been shaped by the role different stakeholders play in the field and the influence they exert on it. Their views not only affect the way the IS curriculum has evolved but also the way young people entering the world of IS transition from academia to industry. Given the importance of the curriculum to the identity of IS, the 'esoteric' crises about its philosophical positioning [24, 25], and the 'applied' crisis caused by inadequate levels of student and professional recruitment [2, 26], it is understandable why researchers study the IS curriculum from a number of different stakeholder perspectives. As [27] explained, the IS discipline is partly defined by its curriculum as represented by two of its main internal stakeholder: (IS faculty, students and graduates).

As stated earlier, studies in the area of IS curriculum mapping are empirical, either utilising conventional survey methods to collect relevant course data, or direct survey approaches that make use of available material which exists online or in printed format. A review of the important work in this area is necessary to ensure that the most appropriate curriculum survey method is adopted when attempting to revise IS curricula. The studies examined employ traditional data gathering methods such as questionnaires [28], direct survey methods [7], or sometimes a mixture of more than one method [29]. Despite their small differences, each method has an important role to play in ensuring that the objectives of a given study a met successfully.

One of the earliest IS curriculum surveys took place between 1977 and 1979. It used a descriptive analysis approach to gather information about the provision of IS courses in the US and the professional careers open to graduates [30]. The purpose of the study was mainly exploratory as opposed to trying to address a specific issue regarding the field. Nunamaker surveyed university catalogues and brochures, obtained through pre-internet era means, such as writing information request letters and placing calls to request materials. This type of data gathering constituted one of the earlier forms of direct surveys which deals with examining relevant aspects of information found in printed materials.

The proliferation and increased popularity of model curriculum recommendations in the 1990s published by ACM and AIS, along with the wide availability of detailed curriculum information on the web, caused renewed interest in the study of the IS curriculum. At the same time, content analysis software and more traditional statistical analysis packages were becoming more widespread, causing a significant reduction in the effort needed to process large quantities of curriculum data. In one such study which unusually focused on postgraduate IS courses, [31] considered the status of Master's degree programmes in the US by examining the commonality of topics covered and their adherence to IS '97: Model Curriculum and Guidelines for Undergraduate Degree Programs in Information Systems [32]. The absence of a comprehensive postgraduate model curriculum at the time, explains the reason behind the authors' decision to use IS'97 which had been designed to cover the undergraduate curriculum. Noticeable research, which made use of the postgraduate curriculum recommendations were published five years later by ACM and AIS [33] and utilised a direct survey approach of gathering course data from university websites [34].

Yang's findings were based on a survey of 273 universities offering postgraduate courses in IS. They showed that the majority of the programmes demonstrated close adherence to MSIS 2006. In another case examining undergraduate courses, [35] gathered and analysed data from university websites for a study in computing education that aimed to distinguish between the different disciplines within the computing field and the academic programmes that define it. More specific direct surveys on the IS curriculum followed after the publication of IS 2002 [36]. Some of these studies relied on the IS 2002 curriculum recommendations to examine the level of alignment of IS courses to the model, thus, measuring the popularity of subjects available to undergraduate students [37-39]. Similar empirical research methods were also used to conduct IS curriculum surveys that examined the correlation between graduate level skills and the skills expectations of industry [13, 40, 41]. For the majority of these studies, as is the case with [6, 42], data was collected from relevant websites to ensure the most up-to-date information is used.

The curriculum surveys which were carried out in the 2000s coincided with a noticeable rise in the accreditation of IS courses by AACSB and ABET [43, 44]. As a result, researchers looking to accurately identify IS course samples for their studies, often chose to include courses which were listed as accredited by the aforementioned organisations. Many of these researchers were influenced by the early work of [45] who set out to profile the IS curriculum of 108 accredited courses. Adopting the same direct survey approach, the authors of that work showed that the rapid technological changes in IT necessitate the frequent revision of the IS curriculum if it is to remain relevant to the needs of industry. Assuming a similar reasoning about the importance of the relevance of the IS curriculum, [38] used a direct survey method to study over 400 accredited IS courses by examining their adherence to the IS 2002 curriculum recommendations. By accurately mapping the reviewed curriculum to IS 2002, the authors strove to support the debate about curriculum modernisation and the skills alignment with industry requirements. With similar objectives, [42] conducted a comparable survey which analysed the IS provision from 240 universities. Although this study did not exclusively include accredited courses, it had the same overall aim of supporting the debate about curriculum and skills relevance at a time when IS student enrolments were falling. In order to improve the accuracy of their analysis, the authors of that study chose to use the draft IS 2010 curriculum recommendations (known as IS 2009 at the time) as the basis for their analysis. More recently, one of the first studies to make use of the latest curriculum recommendations, IS 2010: Curriculum guidelines for undergraduate degree programs in Information Systems [46], is found in the work of [29] who examined 127 AACSB accredited IS courses in an attempt to evaluate how well the IS 2010 recommendations have been adopted by programmes in the US.

[7] encapsulated all of the important characteristics found in IS curriculum survey studies discussed so far. Methodologically, this study clearly demonstrated the value and validity of direct survey as a method capable of accurately collecting secondary data that exists in printed or online format. At the same time, it showed the importance of using an appropriate data sample selection method (in this case partly using AACSB accredited courses) in order to ensure the reliability and validity of the data set. Finally, it used one of the model curriculum recommendations (IS 2002) as a framework for structuring and analysing the data, and presenting the overall results.

4   Applying Direct Surveys to Curriculum Mappings

Curriculum mappings using direct survey approaches are not synonymous with the conventional meaning of content analysis, although sometimes the terms are used interchangeably. Early curriculum surveys relied on the acquisition of relevant data through traditional survey techniques such as questionnaires and interviews. [45] provide one of the earliest curriculum survey examples which involved the distribution of questionnaires. Only a few years later, [47] carried out a similar study aiming to understand the changes in the IS curriculum over a period of study. Their method was also based on a survey questionnaire which was administered to a relatively large number of IS academics.

The expansion of the web saw the publication of extensive course information on university websites. As a result the web, as a new communication medium, shifted the data collection research methods away from questionnaires that necessitated interaction with humans, to more unobtrusive collection and data parsing processes which often benefited from software tools that performed data coding. [6] explained their direct survey method, which underpinned a comparative study of approximately 900 institutions offering IS courses, as a survey of university websites to collect relevant data. In a similar way, [38] made a case for online course surveys by arguing that web sources contained more accurate information than printed catalogues which are updated infrequently. [42] utilised exactly the same web-based data collection approach, as did [7] who explained its value by arguing that the collection of data in a systematic way facilitates its standard quantification.

Two further cases of direct survey of IS curriculum characteristics are found in [48, 49] who carried out two complimentary surveys into the proliferation of career tracks in the design of IS courses. In each case the authors investigated course content available on university websites. [50] discussed the content of business information modules which were part of accredited library and information studies programme, as part of an investigation which involved surveying 48 programmes. The methodological approach used by White was very similar to those described earlier: most content was obtained from webpages, with some unavailable material acquired through staff requests via email. In another study, [51] investigated the role of ethics in the business curriculum by performing two related studies. The first considered data gathered from websites, while the second analysed the content of text books. In a final example, [29] investigated the adoption of IS 2010 by undergraduate degree programmes in order to ascertain their compliance with the new guidelines. Their investigation was supported by two supplementary surveys. In the first instance, the authors used a direct survey to collect data directly from programme catalogues available on university websites. The second survey necessitated additional data which was acquired though telephone interviews. Although rare nowadays, data collection methods such as the latter, or a survey distributed via email [52], still exist as supplementary means of obtaining course data that is not available online.

5   Factors to Consider When Undertaking IS Curriculum Surveys

The majority of the curriculum survey research reviewed in the previous section exhibits three main characteristics that describe the way they were structured: study design, data collection and results. In most cases, the first two characteristics appear under the heading of 'method', 'methodology' or some other synonymous categorisation [7, 38, 42]. From a methodological standpoint, the validity of a study of this type can be ascertained by examining the variables, sampling and coding scheme used.

5.1   Variables

Selecting the appropriate variables, which theoretically determine the outcome of a study as a result of recording information about data items which can be contextually different, is predetermined by the type of course under review. The welcome restriction imposed by variables significantly reduces the margin of error in the design of such studies. This can be explained by considering the work of [47] whose questionnaire featured 120 questions, covering areas about the institution which was being surveyed and the courses it offered. Their approach can be characterised as analytical, as it was trying to determine descriptive information about courses, modules and the teaching software tools used to deliver the curriculum. As such, there are no predictions, generalisations, hypotheses or complex extrapolations about relationships between the data. Instead, the scope of the analysis is determined by the level of granularity of the questions used, and the results are primarily determined by univariate or bivariate analysis. Information derived from questionnaires of this type is no different to the information that is extracted from the course catalogues available on university websites. A typical direct curriculum survey, such as the one carried out by [7], is governed by the pre-existing information that is available online. A small variation to this conventional approach can be seen in [29] with the introduction of variables which calculated scores based on the occurrence of certain modules categories within courses. Even this small variation, however, does not constitute a departure from the standard approach of measuring the curriculum to determine their offerings.

5.2   Sampling

In addition to variables which are crucial for determining what IS curriculum information is to be recorded and analysed, sampling is another important aspect that can influence the overall findings of a study by determining the number and type of courses included. In most cases, the size of the sample is constrained by the objectives of the study. The type of sample used, such as AACBS accredited courses or courses offered by universities in a particular geographical area, characterises the scope of the study, and by implication, the significance of its findings. As an example, [37] used a sample of one, as part of a case study about the IS curriculum in a single institution. At the same time, [6, 45] examined hundreds of IS courses which were qualified as AACBS accredited, while excluding everything else. In another example, [7] sampled general four-year long IS courses found across the US without taking into account their accreditation status. Ultimately, the size of the sample determines the degree of generalisability of the results, while the type affects data homogeneity.

5.3   Coding Scheme

Whereas variables determine the data units being recorded and samples frame the scope of the direct survey data, coding schemes characterise the nature of surveys. Coding schemes can range from simple to detailed. An example of a simple coding scheme can be seen in the work carried out by [30]. In this survey, the author measured the provision of IS courses (mixture of accredited and non-accredited courses), and analysed their names. Similarly, [45] carried out a relatively simple study by coding the modules offered by the AACBS accredited universities surveyed, and ranked them to demonstrate their popularity. In both these cases the information recorded as part of the variables used was of limited depth, since it did not contain details such as the structure of modules, credit weighting or any software tools used in the curriculum. Lack of such detail does not negate the value of the work of either of the aforementioned authors. Their work simply offers a narrower perspective of the curriculum. A more detailed coding scheme was used by [47] who not only considered the type of data seen in previous studies, but also elaborated on specific aspects of courses, offering findings with significantly more depth. Specifically, [47] in addition to deconstructing the nature of the courses under examination, analysed the popularity of software tools used to teach programming, while also considering the different operating systems used in the IS syllabi. It is, therefore, apparent that the increase in the amount and depth of data extracted is directly related to the depth of the coding scheme used. Studies by [29, 38, 42] can be seen as having an even more detailed coding scheme by utilising the structures of curriculum recommendations to contextualise their findings. In the case of [38], data collected from different type of accredited IS courses was mapped to the structure of IS 2002, thus showing the adoption levels of IS 2002, and by implication the level of compliance with its recommendations. More importantly, coding schemes which enable the data of direct course survey findings to be mapped to model curricula, increase the contextualisation and validity of the findings. 'Context' and 'validity' in this case does not mean that curriculum recommendation models should be accepted as the definitive blueprints for designing IS courses. Instead, they can be seen as providing a well-recognised and well-defined structure, which despite its shortcomings, enhances the accuracy of curriculum mappings by providing a reference point of comparison.

6   Summary

The work presented in this paper focuses on the relevant research approach capable of supporting empirical survey studies regarding the classification of IS curricula. The paper considers the research approaches relevant to conducting curriculum surveys by reviewing the relevant literature. IS curriculum surveys in the past followed traditional data gathering methods, such as interviews and questionnaires, while more recent web based content analysis has become an important technique. In order to support research in this important area this paper reviews the main techniques used for the benefit of future researchers engaging in similar work. The detailed literature review regarding the design and analysis of curriculum surveys analysed the variety of data collection methods available, while the discussion considered the advantages and disadvantages of each approach.

References

1. Hirschheim, R., Klein, H.K.: Crisis in the IS Field? A Critical Reflection on the State of the Discipline. Journal of the Association for Information Systems 4(1), 237–293 (2003)
2. Hirschheim, R., et al.: Offshoring and its Implications for the Information Systems Discipline: Where Perception Meets Reality. Communications of the Association for Information Systems 20, 1–22 (2007)
3. Lightfoot, J.M.: Fads Versus Fundamentals: The Dilemma for Information Systems Curriculum Design. Journal of Education for Business 75(1), 43–50 (1999)
4. Snoke, R., Underwood, A., Bruce, C.: An Australian View of Generic Attributes Coverage in Undergraduate Programs of Study: An Information Systems Case Study. In: 2002 Annual International Conference of the Higher Education Research and Development Society of Australasia (HERDSA), Perth, Australia (2002)
5. Latham, A.: The Emergence of Undergraduate Information Systems Education in the United Kingdom and its Challenges for the Future: A Pluralistic, Multi-stakeholder Perspective, p. 386. The University of Warwick (2001)
6. Pierson, J.K., Kruck, S.E., Teer, F.: Trends in Names of Undergraduate Computer-Related Majors in AACSB-Accredited Schools of Business in the USA. Journal of Computer Information Systems 49(2), 26–31 (2008)
7. Kung, M., Yang, S.C., Yi, Z.: The Changing Information Systems (IS) Curriculum: A Survey of Undergraduate Programs in the United States. Journal of Education for Business 81(6), 291–300 (2006)
8. Trauth, E.M., Farwell, D.W., Lee, D.: The IS Expectation Gap: Industry Expectations Versus Academic Preparation. MIS Quarterly 17(3), 293–307 (1993)
9. Todd, P.A., McKeen, J.D., Gallupe, R.B.: The Evolution of IS Job Skills - A Content-Analysis of IS Job Advertisments from 1970 To 1990. MIS Quarterly 19(1), 1–27 (1995)
10. Panko, R.R.: IT Employment Prospects: Beyond the Dotcom Bubble. European Journal of Information Systems 17(3), 182–197 (2008)
11. Foster, A.L.: Student Interest in Computer Science Plummets. Chronicle of Higher Education 51(38), A31–A32 (2005)
12. Hirschheim, R.: The Looming Crisis for the IS Field: Where Have All the Students Gone? Wirtschaftsinformatik 49(3), 232–234 (2007)
13. Prabhakar, B., Litecky, C.R., Arnett, K.: IT Skills in a Tough Job Market. Communications of the ACM 48(10), 91–94 (2005)
14. Tuson, A.: University IT Curriculum is Failing UK Enterprises. In: Computer Weekly, p. 12. Reed Business Information Ltd. (2008)
15. Benamati, J.H., Ozdemir, Z.D., Smith, H.J.: Aligning Undergraduate IS Curricula With Industry Needs. Communications of the ACM 53(3), 152–156 (2010)
16. Hirschheim, R., Klein, H.K.: A Glorious and Not-So-Short History of the Information Systems Field. Journal of the Association for Information Systems 13(4), 188–235 (2012)
17. Avison, D., Elliot, S.: Scoping the Discipline of Information Systems. In: Information Systems: The State of the Field, pp. 3–18 (2006)
18. Barki, H., Rivard, S., Talbot, J.: A Keyword Classification Scheme for IS Research Literature: An Update. MIS Quarterly 17(2), 209–226 (1993)

19. Chen, W., Hirschheim, R.: A Paradigmatic and Methodological Examination of Information Systems Research from 1991 to 2001. Information Systems Journal 14(3), 197–235 (2004)
20. Orlikowski, W.J., Baroudi, J.J.: Studying Information Technology in Organizations: Research Approaches and Assumptions. Information Systems Research 2(1), 1–28 (1991)
21. Dwivedi, Y.K., Kuljis, J.: Profile of IS Research Published in the European Journal of Information Systems. European Journal of Information Systems 17(6), 678–693 (2008)
22. Palvia, P., Pinjani, P., Sibley, E.H.: A Profile of Information Systems Research Published in Information & Management. Information & Management 44(1), 1–11 (2007)
23. Avison, D.K., Kasper, G.M., Pernici, B., Ramos, I., Roode, D. (eds.): Advances in Information Systems Research, Education and Practice. IFIP, vol. 274. Springer, Heidelberg (2008)
24. Benbasat, I., Weber, R.: Research Commentary: Rethinking "Diversity" in Information Systems Research. Information Systems Research 7(4), 389–399 (1996)
25. Benbasat, I., Zmud, R.W.: Further Reflections on the Identity Crisis. In: Information Systems: The State of the Field, pp. 300–306. John Wiley & Sons, Ltd., Chichester (2006)
26. Wilson, N., Avison, D.: Double Jeopardy: The Crises in Information Systems an Australian. In: An Australian Perspective in 18th Australasian Conference on Information Systems 2007, Toowoomba, Australia (2007)
27. Sidorova, A., Harden, G.: Achieving Alignment between IS Research and IS Curriculum: towards Stronger IS Discipline Identity. In: AMCIS 2012 (2012)
28. Riihijarvi, J., Iivari, J.: The Practical Relevance of IT Education: Skill Requirements and Education Expectations of Practitioners. In: Avison, D., Kasper, G.M., Pernici, B., Ramos, I., Roode, D. (eds.) IFIP TC 8 Information Systems Conference held at the 20th World Computer Congress. IFIP, vol. 274, pp. 1–13. Springer, Boston (2008)
29. Bell, C., Mills, R., Fadel, K.: An Analysis of Undergraduate Information Systems Curricula: Adoption of the IS 2010 Curriculum Guidelines. Communications of the Association for Information Systems 32(1), 73–94 (2013)
30. Nunamaker, J.F.J.: Educational Programs in Information Systems: A Report of the ACM Curriculum Committee on Information Systems. Communications of the ACM 24(3), 124–133 (1981)
31. Gorgone, J.T., Lidtke, D.K., Feinstein, D.: Status of Information Systems Accreditation. In: Proceedings of the Thirty-second SIGCSE Technical Symposium on Computer Science Education, pp. 421–422. ACM, Charlotte (2001)
32. Davis, G.B., et al.: IS '97: Model Curriculum and Guidelines for Undergraduate Degree Programs in Information Systems. SIGMIS Database 28(1), 101–194 (1996)
33. Gorgone, J.T., et al.: MSIS 2006: Model Curriculum and Guidelines for Graduate Degree Programs in Information Systems. SIGCSE Bull. 38(2), 121–196 (2006)
34. Yang, S.C.: The Master's Program in Information Systems (IS): A Survey of Core Curriculums of U.S. Institutions. Journal of Education for Business 87(4), 206–213 (2012)
35. Anthony, E.: Computing Education in Academia: Toward Differentiating the Disciplines. In: Proceedings of the 4th Conference on Information Technology Curriculum, pp. 1–8. ACM, Lafayette (2003)
36. Gorgone, J.T., et al.: IS 2002 Model Curriculum and Guidelines for Undergraduate Degree Programs in Information Systems. Communications of the Association for Information Systems 11(1), 1–63 (2002)
37. Dwyer, C., Knapp, C.A.: How Useful is IS 2002? A Case Study Applying the Model Curriculum. Journal of Information Systems Education 15(4), 409–416 (2004)

38. Lifer, J.D., Parsons, K., Miller, R.E.: A Comparison of Information Systems Programs at AACSB and ACBSP Schools in Relation to IS 2002 Model Curricula. Journal of Information Systems Education 20(4), 469–476 (2009)
39. Williams, C., Pomykalski, J.: Comparing Current IS Curricula to the IS 2002 Model Curriculum. Information Systems Education Journal 4(76) (2006)
40. Litecky, C.R., Arnett, K.P., Prabhakar, B.: The Paradox of the Soft Skills Versus Technical Skills in IS Hiring. Journal of Computer Information Systems 45(1), 69–76 (2004)
41. Nelson, H.J., et al.: A Comparative Study of IT/IS Job Skills and Job Definitions. In: SIGMIS-CPR 2007 - Proceedings of the 2007 ACM SIGMIS CPR Conference: The Global Information Technology Workforce, Saint Louis, MO (2007)
42. Apigian, H.C., Gambill, E.S.: Are We Teaching the IS 2009* Model Curriculum? Journal of Information Systems Education 21(4), 411–420 (2010)
43. Hilton, T.S.E., Lo, B.W.N.: IS Accreditation in AACSB Colleges via ABET. Journal of the Association for Information Systems 8(1), 1–U2 (2007)
44. Lidtke, D.K., Yaverbaum, G.J.: Developing Accreditation for Information Systems Education. IT Professional, 41–45 (2003)
45. Maier, J.L., Gambill, S.: CIS/MIS Curriculums in AACSB-Accredited Colleges of Business. Journal of Education for Business 71(6), 329 (1996)
46. Topi, H., et al.: IS 2010: Curriculum Guidelines for Undergraduate Degree Programs in Information Systems. Communications of the Association for Information Systems 26(1), 69 (2010)
47. Gill, T.G., Hu, Q.: The Evolving Undergraduate Information Systems Education: A Survey of U.S. Institutions. Journal of Education for Business 74(5), 289 (1999)
48. Hwang, D., Soe, L.: An Analysis of Career Tracks in the Design of IS Curricula in the U.S. Information Systems Education Journal 8(13), 17 (2010)
49. Soe, L.L., Hwang, D.: Career Track Design in IS Curriculum: A Case Study. Information Systems Education Journal 5(29), 19 (2007)
50. White, G.W.: Business Information Courses in LIS Programs: A Content Analysis. Journal of Business & Finance Librarianship 10(2), 3–15 (2004)
51. Stone, G.W., et al.: A Content Analysis on the Role of Ethics in the Business Curriculum. Journal for Advancement of Marketing Education 9, 31–42 (2006)
52. Ryker, R., Fanguy, R., Legendre, A.M.Y.: Undergraduate Special Topics Courses: What's on the Menu? Journal of Computer Information Systems 49(2), 81–85 (2008)

Strategic Implementation of "professional Massive Open Online Courses" (pMOOCs) as an Innovative Format for Transparent Part-Time Studying

Rolf Granow, Andreas Dörich, and Farina Steinert

Lübeck University of Applied Sciences, Mönkhofer Weg 239, 23562 Lübeck, Germany
{rolf.granow,andreas.doerich,farina.steinert}@fh-luebeck.de

Abstract. Lübeck University of Applied Sciences has a strong profile in occupational online degree courses and scientific online training. The university will expand its strategy of opening up in the following years by "Massive Open Online Courses" for working people (pMOOCs) in order to allow people a development through education.

The University explores now systematically in a multi-year research project how the freely accessible and highly scalable online courses can be developed and embedded on a permanent basis through integration as standard offers. Target groups of pMOOCs are people without a University degree who wish to develop on a Bachelor level with the option of an academic degree as well as people who are interested in a postgraduate training on a Masters level.

With this range, it will be possible to determine those factors that effectively convey the permeability of the higher education system to the public by integrating pMOOCs.

Keywords: massive open online courses, online study programs, online education, e-learning, university development.

1 University Opening by MOOCs - Background Information

Starting from the U.S. recently so-called Massive Open Online Courses (MOOCs) are offered on a large scale. Everybody who has access to the internet also has access to (usually) free courses at university level.

MOOCs use fundamentally the available technologies of Web 2.0, in particular video, interactivity and collaboration opportunities offered by social networks.

The current MMB Trend Monitor, which examines the developments in the e-learning market regularly, classifies MOOCs in addition to Mobile Learning and Social Learning as the most important future trend of learning in the next three years [1]. The NMC Horizon Report, which deals specifically with the proliferation of technology in higher education, comes to a similar conclusion with regard to the relevance of MOOCs [2].

Despite all the hype about high numbers of participants in university MOOCs first empirical studies have shown that it cannot be assumed that an automatic transition to

S. Wrycza (Ed.): SIGSAND/PLAIS EuroSymposium 2014, LNBIP 193, pp. 12–25, 2014.

the formal system of higher education will take place through the course participation. It turns out that the transition rate for the purposes of credit point acquisition is currently extremely low [3].

Accordingly, the participants also rarely achieve an academic degree this way. The MOOC research attributes this phenomenon to the primarily intrinsically motivated learners (experimentation, curiosity, etc.) [4]. This in turn has implications for the design of appropriate operating models for the MOOC format at universities.

MOOCs previously presented themselves in primarily two different types [5]: xMOOCs and cMOOCs. In xMOOCs ("x" = extension), the focus is in a clearly structured, predominantly presentation, arranged with video material. The learning outcomes are in general clearly defined and the didactic design is created accordingly. In addition to testing and forums as part of more complex types of tasks, for example essays and peer-review processes are used.

Depending on the topic and the focus on discursive exchange, orientation of the course towards cMOOCs may be appropriate ("c" = connectivism). Here the focus is on a common "connected" learning process. This is "negotiated" among the participants in discussions and discourses and realized through a variety of communication tools.

Other forms, which are already experimented, for example, are bMOOCs [6] ("b" stands for blended), in which a normally "closed" event will be opened for outside participants, and aMOOCs [7] ("a" stands for adaptive). aMOOCs aim to break up the rather rigid structure of the xMOOCs and to enable a more personalized learning experience. In this case the preferred learning behavior should to be met by various technically realized learning options.

Up to now MOOCs are still insular areas of experimentation, in which the applied teaching and learning concepts are studied, isolated and tested. There are so far only few insights into the organizational and structural embedding of MOOCs in universities. This also applies to MOOC-related capacitive considerations as well as the permeability and the part-time learning. The quality of upscaled MOOCs and quality assurance also raise massive questions [8]. The use of MOOCs as a key structural and profile feature of a university as well as the inherent need to effectively influence and design transitions of the participants, is a largely unexplored area. In particular, the focus in MOOCs has not been placed yet on the potential and effects for special target groups, as would be given in a format specifically designed for working professionals.

2   Massive Open Online Courses as a Strategic Element of Lübeck University of Applied Sciences

Since 2001 Lübeck University of Applied Sciences (LUAS) enables professionals to access to Bachelor's and Master's degrees through their online distance learning courses. Approximately 13% of LUAS students already use the option to complete higher education studies and continue working at the same time - with an increasing tendency.

The university continuously improves the general working conditions for career development, inter alia, access and crediting. The consequent opening of LUAS for new audiences through online distance learning courses are also operated in the joint project "Offene Hochschulen in Schleswig-Holstien: Lernen im Netz, Aufstieg vor Ort" (Open Universities in Schleswig-Holstein: Learning in the Network, Career Development at Home)[1].

Based on its broad expertise in the field of occupational online studies and in the online education LUAS will use MOOCs to expand its opening in the sense of lifelong learning and permeability dynamically [9]. In the first MOOCs in the German higher education (University of Frankfurt) only about 13% of students had logged in, about 75% of the participants were employees or freelancers, mainly with university degrees [10]. In order to illustrate the orientation of MOOCs to professionals, the term "professional Massive Open Online Courses" (pMOOCs) is used which LUAS would like to establish as a high quality, quality assured and reliable profile feature of an open university.

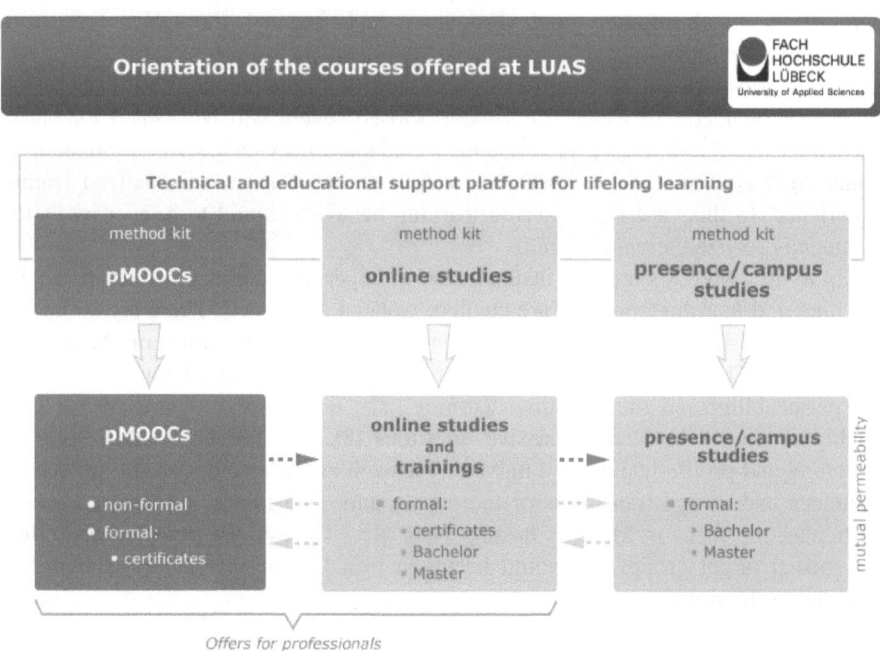

Fig. 1. Alignment of programs offered by the Lübeck University of Applied Sciences

---

[1] This joint project receives funding under the competition "Aufstieg durch Bildung: Offene Hochschulen" from the federal budget as well as funds from the European Social Fund of the European Union (ESF). Funding codes of the project are 160H11057 to 160H11061.

pMOOCs offer great potential because of the openness to acquire a far higher number of professional and other professionally qualified people for scientific education than before. First own experiences with open online courses were made at LUAS in the winter semester 2013/14 with a MOOC on the topic of marketing [11]. This course has achieved a wide audience with its approximately 7,000 participants and offered the possibility of a final formal examination. The broad base of pMOOCs at the university will be systematically analyzed in a federally funded multi-year research project with the involvement of the target groups and represent the professional foundations for a sustainable implementation of the course format.[2]

The previous online courses at LUAS are a closed format. The participants have to overcome formal, organizational and financial barriers to participate in the study or training operation. pMOOCs on the other hand are a format without barriers. Any interested person with internet access can register with minimal effort and participate free of charge. The format is not primarily focused on the acquisition of formal certificates, but also allows access to non-formal learning processes.

The following table shows the paradigms of the two formats in the overview:

Table 1. Comparison of the paradigms of pMOOCs and the previous online studies at Lübeck University of Applied Sciences

pMOOCs	Previous online studies
free access for all	enrollment with access requirements and admission restrictions
open groups	closed groups
no legal relationship with the university	students become members of the university; in continuing education participation contracts are concluded
participation free of charge (fees may only occur later, e. g. for testing)	fee schedules, semester fees, if applicable fees for continuing education
little support (peer review and peer collaboration)	intensive care (binding professorial and mentoring care)
initially non-formal learning has priority; formal learning is possible	formal learning has priority
self-organization among the participants	participants are organized

---

[2] This project receives funding under the competition "Aufstieg durch Bildung: Offene Hochschulen" from the federal budget as well as funds from the European Social Fund of the European Union (ESF). Funding code of the project is OH21017.

The use of pMOOCs as freely accessible educational format for working professionals leads to massive changes in the previously proven teaching and learning. It is therefore the objective to orientate the scenarios of pMOOCs to the specific learning needs of professionals, without jeopardizing the paradigms of the format. But the project will not only focus on the transition rates to formal studies as a success factor. All educational potentials of pMOOCs should rather be explored for an acquisition of competence without the formal construct of a university. In this respect, in the scenarios being developed, the area of non-formal learning is actively considered. pMOOCs should therefore keep open all education and orientation options for the participants.

Since non-formal learning plays an essential role in pMOOCs, the transitions from non-formal and formal learning should be optimally designed, pMOOCs developed permeable to formal study programs and the integration of vocational skills in the learning process should be supported. pMOOCs should open LUAS to a large extent for both working professionals with no university degree (Bachelor level) as well as for university graduates who wish to postgraduate (Master level). Furthermore the conception of relevant study modules that can be used for both non-formal and formal with certificate shall be explored. Appropriate certificates will then be credited to both the existing part-time online degree programs as well as to LUAS presence courses.

Reliable insights should be obtained in the project, to what degree pMOOCs can substitute usefully technical and business-related courses and how to evaluate their acceptance in the target groups of the working population compared to existing part-time study formats.

3   Research Questions and Methodological Approach

With the objective to provide a systematic basis for the strategic implementation of pMOOCs at the LUAS, seven research issues will be processed in the next four years.

1. pMOOCs as an open format require a different orientation with respect to the learning arrangement as the proven online distance learning courses with their closed groups. The question is therefore how MOOCs must be designed in content, didactically and technically in order to specifically address professionals.
2. pMOOCs could support the transition into the formal higher education system. However, the available empirical and theoretical findings on completed MOOCs of universities have indicated low rates of credit point allocation so far. This also presumes a low transition probability in a regular course with an appropriate degree. However, in order to achieve the desired effect, the question to answer is how concepts should look like to increase the transitions of employed pMOOC-participants in the formal higher education system and how they should be designed to achieve this target.
3. pMOOCs are a consistently open format. Therefore, the philosophy based in the present project should be to be open to all visual scenarios of the participants. Therefore, the intention of learning, without the intention of certifying has also to be encouraged, as the formal transition to the higher education system through

transfer of credit points. The question is how non-formal learning skills acquired in pMOOCs can be credited for a transition to the degree program.

4. For a broad and sustained implementation of pMOOCs as further education of the university some basic questions such as the capacity effectiveness, teaching credits, the organizational embedding and quality management have to be clarified. This raises the question of how pMOOCs can be embedded as sustainable further education for professionals as a profile element in the university. It is important to clarify which conditions must be developed at the university at the organizational, capacitive, formal and personal level, and what kind of concepts may be promising.

5. How can pMOOCs be designed in the immediate web-based interaction with their target audiences, customized with demand, so that they still map the academic profile of the university? How can structures of social networks be used to meet an active, bilateral exchange of requirements and needs of the target groups on the one hand and the specific scientific potential and knowledge of the university on the other hand?

6. There are a dynamically growing range of MOOCs. Therefore it makes sense to build on these offers and to concentrate own activities with focus on the development of such pMOOCs in which the university can identify their particular skills. This leads to the question: How can existing MOOCs of other education institutes be integrated value-adding in the concept of pMOOCs?

7. The resources required for the development and the sustainable operation of MOOCs can vary considerably depending on multimediality, support costs and course organization. The challenge is to achieve the best possible cost-benefit ratio. Accordingly the quality requirements of the participants and universities are to be reconciled with the current and future available financial and time resources. The question is in this context, how viable resource calculations for MOOCs various multimedia, didactic and organizational expression may appear in a future-oriented portfolio of an open university. Here you have to start from different forms of the courses that are conceptually explored in research question 1 and highlighted under standardization aspects. With the objective of a coherent resource calculation relevant fixed and variable costs parameters like development and teaching capacities as well as the necessary technical infrastructure will be analyzed. Quantities, economies of scale, learning curves and risks will be considered and appropriate costing models for the higher education sector will be developed and optimized.

The formulated research issues suggest an explorative research approach because as a phenomenon in the higher education context MOOCs are still largely unexplored. Therefore the project will give indications on hypotheses on MOOCs, their participants and their implementation at universities.

LUAS has the aim to consider the MOOC philosophy ("involving all") consequently. This should therefore be reflected in the processing method of the research questions. Therefore it is planned to approach the raised questions mainly in participatory action research [12]. The action research principle works on the philosophy of Web 2.0, it is based on the equal participation of stakeholders. In the research context this means that researchers and research objects reflect the process together, identify causes of problems and successes and develop improvements. The result is a fertile cycle of action and reflection [13]. However, the target groups must be willing to participate. In MOOCs this condition is met as very own feature.

In the project the participant observation as a classic element of action research, the standardized and semi-standardized interview, workshops, standardized survey (online) and additionally the secondary research are used. The planned competence measurement will use more quantitative, standardized instruments than existing, proven questionnaires. Objective is a multi-step evaluation process that integrates subjective evaluations of the participants.

4   Work Planning

The project is divided into six work packages, of which the first four deal directly with the design, development, testing and evaluation of the individual pMOOCs made on the basis of research questions: the research questions are currently being processed on the basis of exemplary pMOOCs. The work package 5 takes over the central medial transposition of the pMOOCs, the 6th work package project management.

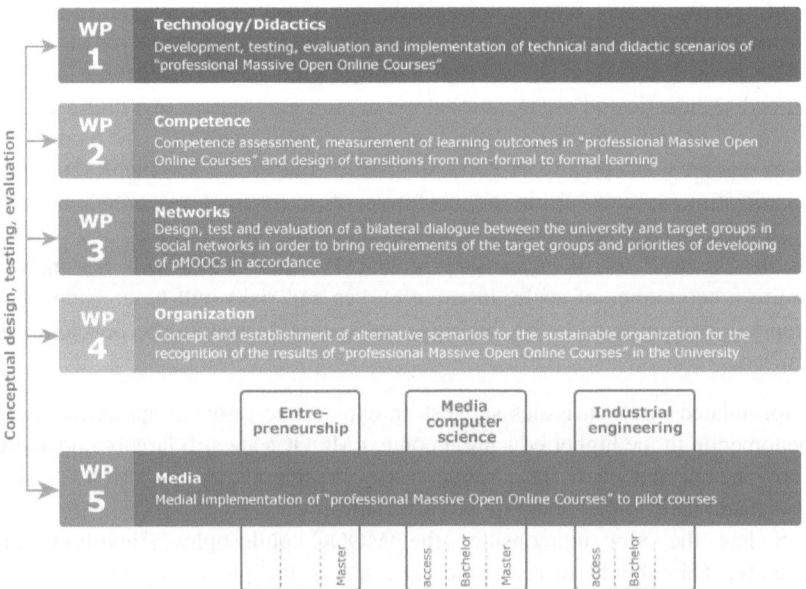

Fig. 2. Representation of the work packages in the first funding phase

In the project six pMOOCs are created. The pilot pMOOCs will be an adequate selection for the research questions for the three disciplines of entrepreneurship, industrial engineering as well as media computer science, sequentially implemented with multimedia and increasing insight. The resulting six pMOOCs will be successively tested and fed back into the development of subsequent courses with regard to the validation of knowledge gained from the work packages 1-4.

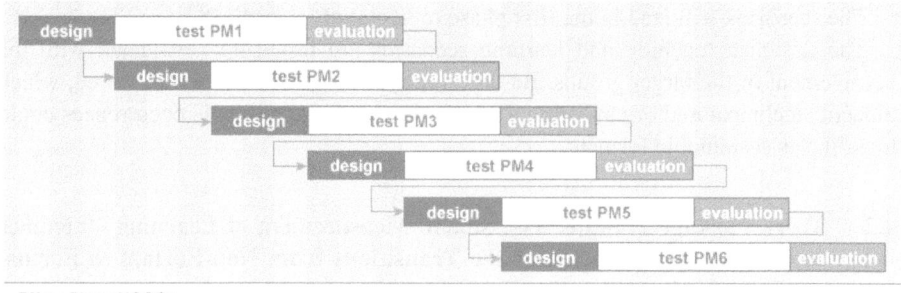

PMn = Pilot-pMOOCn

Fig. 3. The feedback in the reaction of pMOOCs

The selection of the disciplines is due to the thematic priorities, in which the university has appropriate continuing education courses: Entrepreneurship, media, computer science and industrial engineering. There are six pMOOCs to be developed with different scenarios and concepts. The permeability of pMOOC in studies up to master for part-time students will be enabled exemplarily.

4.1   Work Package 1: Development, Testing, Evaluation and Implementation of Technical and Didactic Scenarios of "professional Massive Open Online Courses"

It is important to develop and test didactic and technical formats for pMOOCs, specifically oriented to the specific learning needs of professional, without challenging the paradigms of the format. The scenarios should aim to integrate existing skills of learners in the learning process and to promote the learning together in open groups. Especially xMOOC and cMOOC concepts should thereby be analyzed. With the further opening of the university to the area of undergraduate degree courses, most xMOOC concepts are used, which can also be extended by discursive methods.

It is an essential research motivation to find out which are the different didactic scenarios with good results. In the postgraduate area the focus is of the didactic design is on more elaborate learning outcomes that require correspondingly more advanced teaching methods. Here the research focus will be more on cMOOC elements.

All scenarios in this non-formal settings combine elements of self-directed learning in the form of video-based course materials with social learning settings such as forums discussions and peer reviews. Mainly the development of practical skills and personal skills of the learners are encouraged to secure their employability permanently and to enable advancement through education and continue to work at the same time.

A technical environment accompanying the didactic concept is provided that allows the design of courses for specific scenarios efficiently and prepare the content for collaborative learning in pMOOCs. The specific implementation of individual learning units will strongly differ and is oriented in the context of the intended learning outcomes to the different target groups.

The scenarios will lead in the first phase to a total of six pMOOC designs.

The designed teaching and learning scenarios are tested and evaluated with the involvement of the target groups. Based on the results it will be then explored, which didactic, technical and organizational standards of various pMOOC occurrences could look like as sustainable formats.

4.2   Work Package 2: Skills Assessment, Measurement of Learning Outcomes in pMOOCs and Design of the Transitions from Non-Formal to Formal Learning by Crediting Scenarios

The learning results or the entire formal qualification is classically understood as acquired knowledge. Today the results of the learning process are merged under the broader term of competence [14].

The different components of competences can be received in several ways: Structured learning in educational institutions is allocated to the formal learning, systematic and targeted learning outside of these institutions defined as non-formal learning and we summarize the incidental learning in everyday use, business and leisure with the term of informal learning [15].

The planned format in this context interrupts the traditional assignment of scientific training, since large parts of the competence acquisition take place outside the control of the university and teachers, but are nevertheless important for the outcome of the course.

In the project appropriate forms of competence assessment and the measurement of learning outcomes are examined for their suitability for part-time learners in pMOOCs. This is important in order to ensure and demonstrate the high quality of learning offers. Certificates of higher education have only skills that have been acquired in formal courses of study or training courses. For use in pMOOCs this concept has to be extended so that non-formal learning outcomes can be mapped according to quality standards and acquired skills can be verified.

It is focused also how non-formal skills can be credited to the online degree programs of the university. It is shown in a model on the basis of previous research results and the current legislation how crediting as a lump sum, individual or combined sum from the open format of the pMOOCs can take place in courses. For model development thereby the specific development of the participants' skills of pMOOCs is used. To determine the skills acquired didactically useful examination forms are designed and tested, which go far beyond the knowledge query with multiple-choice tests and allow a secure competence assessment. On this basis pMOOCs are then eligible on study courses.

Objective is a quality assured credit on the basis of a valid transparent process, which is accepted by the participants and that meets the quality criteria of quality, equivalence, accountability, simplicity, sustainability and transferability requirements. Within this process it is paid attention to appropriate descriptors for skills and learning outcomes how also specified by the DQR for a comparable basis [17]. The required resources and appropriate funding models also need to be carefully taken into account.

With the objective of identification of model, solid credit scenarios learning outcome descriptions are exemplarily documented, tentatively performed equivalence tests, explored the required formal establishment within the university and prepared outlines methods of information and consultation both internally and externally as well as prepared the structures for a formative evaluation[3].

In addition to the development of adequate imputation procedure you also have to explore carefully during the pMOOC development the contribution to the skills development they can afford for a transition into the formal higher education system but without just copying the format of regular, closed course. The objective is to emerge independent pMOOC formats with connectivity to the formal system.

4.3   Work Package 3: Design, Test and Evaluation of a Bilateral Dialogue between the University and Target Groups in Social Networks in order to Bring in line Requirements of the Target Groups and Priorities of Developing pMOOCs.

pMOOCs allow a new port and involvement of stakeholders. Social networks enable a dialogue with the target groups which no longer need to take place via multipliers and market research tools.

In an active, bilateral social community management, the target groups decide which pMOOCs should be offered in what format. For this purpose, appropriate strategies and instruments have to be designed and tested. For a large range of coverage existing channels are used, which play a role in Germany and Europe for the operation and marketing of MOOCs. Beyond the specific offer participants can actively participate in the learning scenario and distribute trans-medial. The social networks are therefore directly involved as multipliers in the efforts of the range extension. An organic growth of the number of participating, i.e. such growth, based on recommendations of the offer in their own environment, offers high potential to increase the sustainability of pMOOCs. To be emphasized is the characteristic of social networks, to address those interested subject-specific targeted and engage in a collaborative communication process. This contributes to a needs-based approach in the design of specialized learning opportunities in a particular way.

---

[3] Scientific support of the BMBF initiative „Anrechnung beruflicher Kompetenzen auf Hochschulstudiengänge" (ANKOM), HIS Hochschul-Informations-System GmbH, Institut für Innovation und Technik (iit) der VDI/VDE Innovation und Technik GmbH (ed.) (2010): Anrechnungsleitlinie - Leitlinie für die Qualitätssicherung von Verfahren zur Anrechnung beruflicher und außerhochschulischer erworbener Kompetenzen auf Hochschulstudiengänge, refer to: http://ankom.his.de/know_how/ anrechnung/pdf_archiv/ANKOM_Leitlinie_1_2010.pdf (date of last access: 16.5.2014).

4.4   Work Package 4: Concept of Alternative Scenarios for the Sustainable Organization and to Sustained Operation of pMOOCs at the University

So far MOOCs are carried out at universities on pilot-scale, sustainable financial or organizational implementation do not take place. Therefore, a major focus is on developing concepts that integrate pMOOCs permanently in the university operation and its resources. Objective is to create a business model that ensures adequate human, technical and infrastructural resources for sustainable operation of the offers. So far, no model was able to meet all the items referred to without also compromising the openness resulting of the free courses. It will be important to achieve economies of scale, so that created technical and didactic basics are frequently and durable used, and the idea of pMOOCs is carried in the entire university.

One important approach is to be examined, to represent degree courses in the long term by pMOOCs and thereby open up effective capacity resources. In this context, employment law issues of teaching credits and capacity efficiency, as well as the rights of pMOOCs have to be clarified. It should also be examined in this context in which extent pMOOCs can be integrated into the classification of performance bonuses for professors, considering the W salary scale introduced since 2005. Financing options through examination fees, registration fees, or third parties such as companies in the private sector have to be considered. Within the project Collaborations with other universities and training providers as potentially promising concepts for pMOOCs will be explored.

Ultimately concepts for the sustainable implementation are designed and exemplary tested for the level of the operational course organization in this work package.

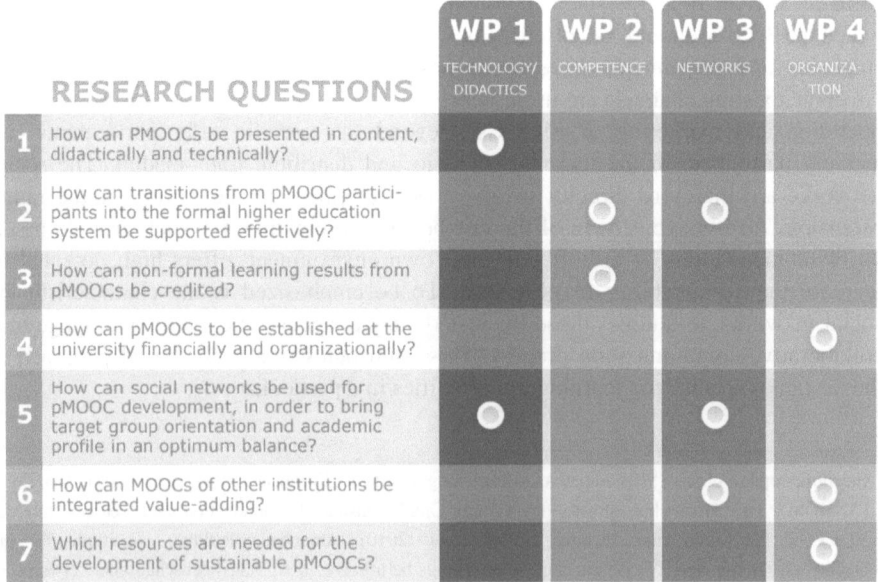

Fig. 4. Illustration of the connection between research questions and work packages 1 to 4

The following chart illustrates the interlocking of the work packages 1 to 4 with the developed research questions.

4.5 Work package 5: Multimedia Implementation of "professional Massive Open Online Courses"

The concepts to technology and didactics, competence assessment, accreditation, network management and organization to be developed in the work packages 1 to 4 cannot be seen in isolation of an exemplary implementation. In order to enable a scientific review of the concepts of all work packages integrally, while ensuring the strategic implementation of pMOOCs as part of the curriculum of LUAS, pMOOCs will be implemented on a pilot basis during the first phase in initially three topics medially.

A major focus in the media production will be on future-proof video formats. They will be developed with interactive elements that require modern cutting technologies and include outdoor scenes video scenarios. Green-screen technology, screencast technology and tablet recording will be used as well as interview formats that have a personal and involving character. Simulations, animations, audio elements, photos and graphics will also be integrated.

The main focus is also on the intelligent integration of media-based interactive services, task types and motivation-enhancing gamification elements such as Open Badges or interim certificates. Objective is to produce media based the optimal balance between professionalism, clarity and entertainment value [16].

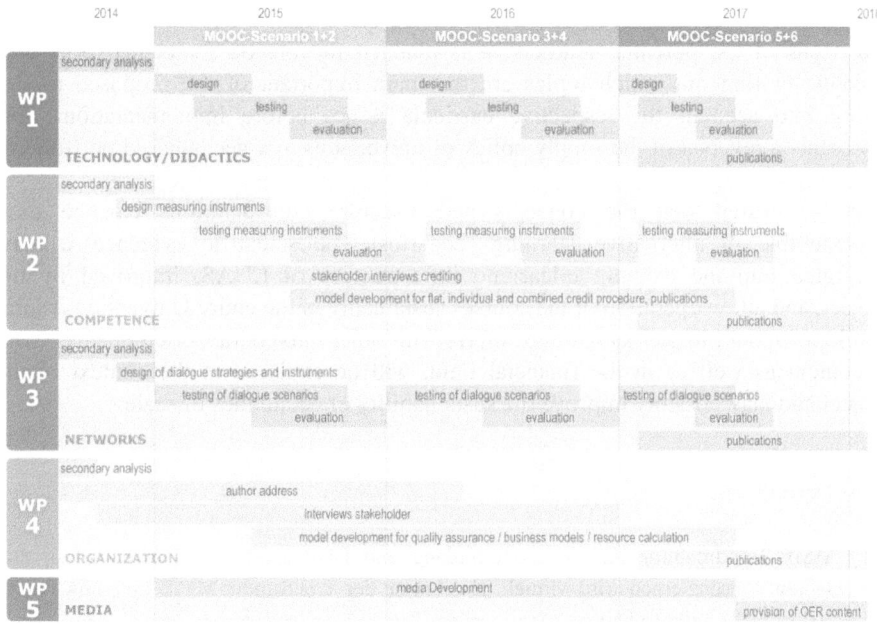

Fig. 5. Illustration of the project's progress with milestones

5   Approach to Sustainability

The implementation of pMOOCs will then be a lasting impact for the opening of the university, if these offers will also be available for long term and the infrastructure which has to be generated will be stable and expandable in diversity.

It is intended to convert the resulting pMOOCs on Bachelor and Master level after the project end in 2018 into capacity effective standard offers of LUAS under existing programs. This means a long-term financing of pMOOCs after the project period from the university budget.

Budget-financed courses from the regular university programs will be substituted later by pMOOCs accordingly. After the end of project funding the teaching staff from the university will work for the support of the pMOOCs. This expression emphasizes the permanent implementation of the pMOOCs and also creates a resource base for the necessary continuous review and updating of courses. The project creates the necessary formal and organizational requirements while enabling entanglement of pMOOCs with the other offers of LUAS. Such implementation requires obligatory legal and organizational arrangements with the faculties of the university, with its presidency, as well as with the administration.

The creation of an incentive to promote permeability and advancement through education about the target agreement with the state of Schleswig-Holstein which supports the funding of higher education can be a supporting element to foster the pMOOC implementation.

Operation, support and creation of pMOOCs with the objective of familiarization of participants in the university are sustainable, if they help to increase the demand for study places in the long term. In order to ensure that the offerings reflect the interests and needs of the potential learners, it is planned to vote on social networks in a dialogue to find out which topics are the main important of the proposed topics, taking into account the respective curricula issues before implementation. This represents a reversal of the supply policy of universities to a needs-based portfolio of courses.

It is ensured that the concepts and structure considerations can be used interoperable in alternative learning scenarios. The scientific evidence will be integrated into the existing e-learning infrastructure of LUAS, improved by the project, and allow operation of the courses sustainably in the entire University. A joint development and implementation of offers with other universities or companies, both in content as well as in the financial field, will be explored in the context of the project and may possibly improve the sustainability of economies of scale.

References

[1] MMB-Trendmonitor I/2013. Weiterbildung und Digitales Lernen heute und in drei Jahren. Präsenzlernen wird virtuell. Ergebnisse der Trendstudie MMB Learning Delphi 2013 (2013), refer to http://www.mmb-institut.de/monitore/trendmonitor.html (date of last access: May 16, 2014)

[2] NMC-Horizon Report, 2013 Higher Education Edition (2013), refer to `http://www.nmc.org/pdf/2013-horizon-report-HE-DE.pdf` (date of last access: May 16, 2014)

[3] Kolowich, S.: A University's Offer of Credit for a MOOC Gets No Takers. The Chronicle of Higher Education (July 08, 2013), refer to `http://chronicle.com/article/A-Universitys-Offer-of-Credit/140131/` (date of last access: May 16, 2014)

[4] Meinel, C., Willems, C.: openHPI – Das MOOC-Angebot des Hasso-Plattner-Instituts, Technische Berichte des Hasso-Plattner-Instituts für Softwaresystemtechnik an der Universität Potsdam, 79 Universitätsverlag Potsdam, Potsdam, p. 8 (2013)

[5] Bremer, C., Thillosen, A.: Der deutschsprachige Open Online Course OPCO12. In: Bremer, C., Krömker, D. (eds.) E-Learning zwischen Vision und Alltag, Münster/New York, München, Berlin, pp. 15–27 (2013); sowie Landesanstalt für Medien Nordrhein-Westfale (ed.). DIGITALKOMPAKT LfM, MOOCs einfach auf den Punkt gebracht, Düsseldorf (2013)

[6] Ibid., p.17

[7] Sonwalkar, N.: The First Adaptive MOOC: A Case Study on Pedagogy Framework and Scalable Cloud Architecture-Part I. In: MOOCs FORUM, vol. 1, p. 22–29 (September 2013)

[8] Bremer, C., Thillosen, A.: op, p. 17 (2013)

[9] Hanft, A.: Lebenslanges Lernen an Hochschulen – Strukturelle und organisatorische Voraussetzungen. In: Hanft, A., Brinkmann, K. (eds.) Offene Hochschulen – Die Neuausrichtung der Hochschulen auf Lebenslanges Lernen, pp. 21–22. Waxmann, Münster (2012, 2013)

[10] Refer to `http://www.e-teaching.org/lehrszenarien/mooc/` (date of last access: May 16, 2014)

[11] Refer to `https://iversity.org/courses/grundlagen-des-marketing` (date of last access: May 16, 2014)

[12] Altrichter, H., Posch, P.: Lehrerinnen und Lehrer erforschen ihren Unterricht: Unterrichtsentwicklung und Unterrichtsevaluation durch Aktionsforschung, vol. 4, pp. 13–14. Auflage, Klinkhardt, Bad Heilbrunn (2007)

[13] Altrichter, H., Posch, P.: op, pp. 15–17 (2007)

[14] Heyse, V., Erpenbeck, J.: Kompetenztraining. Schäffer-Poeschel Verlag, Stuttgart, S.XII (2009)

[15] Overwien, B.: Stichwort: Informelles Lernen. Zeitschrift für Erziehungswissenschaft 8(3), 339–355 (2005)

[16] Sloane, P.F.E.: Zu den Grundlagen eines Deutschen Qualifikationsrahmens (DQR). Konzeptionen, Kategorien, Konstruktionsprinzipien. Bundesinstitut für Berufsbildung, BIBB, p. 16 (2008)

[17] Meinel, C.: openHPI – das MOOC-Angebot des Hasso-Plattner-Instituts. In: Schulmeister, R. (ed.) MOOCs – Massive Open Online Courses. Offene Bildung oder Geschäftsmodell?, p. 69. Waxmann, Münster (2013)

Key Principles of Reference Model
for Cost Allocation and Profitability Management

Milos Maryska and Petr Doucek

Vysoká škola ekonomická, Praha
milos.maryska@vse.cz, doucek@vse.cz

Abstract. The proposed conceptual model deals with two areas – Cost Allocation and Profitability Management. This paper shows some limitations of the model, its architecture – the individual layers of the model, key principles of cost allocation on which the proposed model is based, and factors which must be taken into account during the development and subsequent implementation of the model. In conclusion there are several ideas for the future development of the reference model.

Keywords: Performance management, business informatics, allocation, profitability.

1 Introduction

Changes in the economy and overall behaviour of companies, as well as society development, have brought about far-reaching changes in internal company management. Pressure for changes is palpable in the area of measuring results and performances at microeconomic level. (Fischer, Novotny, Doucek, 2013)

Measuring results and performances has a very long tradition. Over the last decades there has been rapid development of tools for measuring results and performances with the support of information and communication technologies (ICT). This support is used in measuring (when the necessary data are collected) for the management of entire organizations, as well as their parts in the form of organizational units or project teams (Strizova, 2013).

A common requisite for fulfilling the sense of measuring as a foundation is a system-wide approach and thinking. In the context of system-wide approach and thinking it ensures interconnection of the phenomena being examined from both qualitative and quantitative views in system unity.

This paper aims to present a conceptual reference model for cost allocation and profitability for efficient management of corporate informatics (REMONA) based on the principles of Corporate Performance Management (CPM). An inseparable part of the presentation is holding a scholarly discussion about the presented model to obtain feedback and opinions on its design from the academic community and from end users. The REMONA model is proposed as part of an academic project of the Faculty of Informatics and Statistics at the University of Economics in Prague in association with the companies Profinit, s. r. o. and Lodestone Management Consultants, A.G.

S. Wrycza (Ed.): SIGSAND/PLAIS EuroSymposium 2014, LNBIP 193, pp. 26–35, 2014.

2   State of the Art

Management of the economy in general and corporate informatics in particular is an area which has to be addressed in detail in the context of the management of an entire company. (Chen, 2004) This is corroborated by the fact that a company's investment in its internal ICTs accounts for a significant and ever-growing portion of all investments. (Dimon, 2013) Therefore, it is not possible for the finance not to be managed and the related costs not to be properly allocated to corporate processes and end users and reflected in the company's pricing.

Cost allocation, pricing, and the related profitability is growing in importance, especially in periods when the company. The market and the economy are undergoing a negative economic development. It is in such periods that managers demand accurate, detailed and up-to-date information not only about the company as a whole but also about its individual parts. (Kral, 2010) Key activities and goals according to (Dimon, 2013; Lomerson, Tuten, 2009; Turban, Leidner, McLean, Wetherbe, 2007) include at present:

- Every company tries to get maximum return on each investment and clearly identify, and in many cases calculate, the benefits of investments.
- Companies try to minimize or eliminate activities and processes which do not generate the required value.
- Companies struggle against changing economic conditions.
- Measuring and managing a company as a whole and company informatics as one of its parts is a phenomenon being closely monitored.
- Proving that investments are warranted (for example, in ICT) and proving the achievement of expected or required results.

All activities involved in the identification and keeping records of costs and earnings are closely related to financial and management accounting and its methods, such as ABC (Activity Based Costing). These methods are a key part of an integrated system for managing the performance of a company, the so-called CPM, in which management, modelling and optimization of profitability, including what-if analyses and cost allocation, play the key role.

CPM does not represent only financial and management accounting as its integral part is technological support, which includes Business Intelligence (BI), reporting solutions, and business analytics. Generally speaking, CPM are activities and solutions used by companies aiming to become successful in the competitive struggle. CPM focuses on support for solving managers' key tasks. (Dimon, 2013; Cookins, 2009)

Research done in the last decade (for example Chandler, 2007; Muhammad, 2010; Remenyi, Bannister, Money, 2007) in the world (Maryska, Wagner 2013; Maryska, Novotny, 2013) and in the Czech Republic shows that management and measuring of the performance of a company as a whole and the individual parts, including company informatics, is difficult and requires relatively complicated solutions (Variana, Farrel, Shapiro, 2004; Remenyi, Bannister, Money, 2007).

3   The Model

The proposed model identifies key dimensions and indicators and interconnects them within designed analytical cubes. REMONA is designed to be easily integrated into a company and easily configured, which enables it to quickly tailored to the needs of a specific company.

3.1   Purpose and Meaning of the Model

The aim of the model is to offer a solution to two key corporate tasks, 'cost allocation' and 'profitability management'. This solution is inextricably connected with the tasks of analyses and in particular, what-if analyses.

For both tasks the model comprises key 'Dimensions', 'Metrics', 'Drivers' and 'Activities', which are addressed as part of corporate informatics. Another requirement for the model is the possibility of its rapid and easy adaptation to a specific company in which it will be implemented. This is achieved in the case of the REMONA model by its logic being implemented as much as possible through appropriate links between data cubes and related dimensions.

To get the full picture we should add that in the case of specific companies or specific allocation rules or analyses of profitability we are ready to make required changes directly in the reference model (solution code) and add new findings to the original model through system feedback.

3.2   Requisites and Limitations to Model Design

The proposed model is based on basic requisites, limitations and requirements which must be fulfilled to ensure that REMONA can be easily and quickly implemented in a company. The model design is based on the following:

- The overall design of the model must be a general one so that it can be tailored to the needs of a target organization.
- The proposed model must support easy and quick integration into corporate architecture.
- The model will be created in such a way that modifications can be made primarily through configuration of the system, although it is possible that some functionality may have to be developed to meet specific requirements.
- During the preparation of the model the necessary dimensions and key metrics must be identified for tasks carried out in a given area.

When designing the model and subsequent implementation of the system it is necessary to answer some key questions, which have to be taken into consideration as they affect the preparation of the proposed model:

- What are current and expected main problems in economics and management of development and operation of corporate informatics and what the priorities of the solution?

- Are some of the standard methodologies (ITIL, CobiT) or proprietary methodology or model used in the management of informatics?
- Is the management of corporate informatics based on the management of services and service level agreements?
- What key metrics are required for the management of the economics of the system for corporate informatics? Are any in use at present?
- Is there documentation of the management of corporate informatics and database management from which data can be obtained? Are there data in them that could be used to design and fulfilment of metrics and dimensions?
- Has an analysis been carried out of the level of ripeness of processes of corporate informatics management and what are the results?
- How high a level of detail will be necessary for analytical tasks in the management of economics of corporate informatics?
- How are costs of PI monitored and what is the place of cost analysis in corporate informatics management?

The proposed REMONA model is designed to permit easy and quick adaptation (modification) of the solution according to the character of the answers to these questions by parametrization without high costs of additional alterations.

3.3   Architecture

The architecture can be described from several views with different degree of detail and elements describing the model.

The basic view of the architecture is represented by individual layers integrated in the model. It is a layer of (Figure 1):

- primary data sources,
- data integration – Data Stage (addressing questions of data pumps (ETL) and data quality),
- core of data warehouse and data marts,– partly addressed in REMONA,
- application layer and user interface layer (object of REMONA),
- a metadata layer passes through all the layers which is of key importance for end users as it guarantees a standard language and description of all indicators and attributes which are part of REMONA and the other layers of the company information system.

The architecture of the model shown in the following picture (Figure 1) is based on the traditional architecture of a BI solution and modified for the purposes of the model with the aim of allowing its integration into the architecture of an ordinary organization. The picture shows in detail a view of the individual components of the architecture described above as part of data warehouse and application layer.

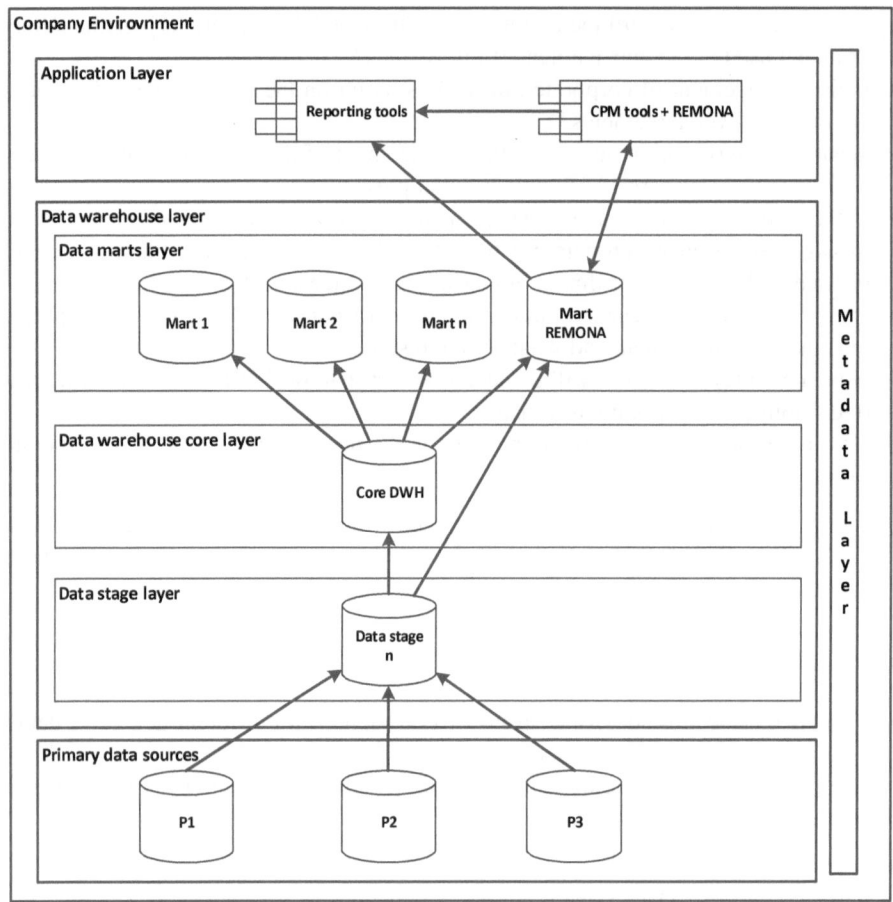

Fig. 1. Architecture of model at company level, Source: Authors

The model is built on top of already consolidated data sources which usually represent a data warehouse in an organization (Core DWH) in 'Data warehouse core layer'. Data are entered into DWH from primary data sources via a data stage (one or more) and may (but need not be) be entered into the data mart in 'Data marts layer'. Loading data from primary systems to DWH and data marts is not the subject of the proposed model.

When designing the model a model of data warehouse/data mart will be created to serve exclusively the needs of the proposed model and meet requirements imposed on it by the proposed solution. The proposed data warehouse is conceived to be as generalized as possible so that it can be implemented in any type of company that has decided to carry out tasks of cost allocation and profitability through company informatics. We start from the assumption that the basic models and principles of cost allocation (based on Activity Based Costing) and profitability management is not affected by the character of the company. The differences stem from methods of parametrization of the created models.

Above the data marts layer (it can be also above the DWH layer) is a layer of application tools ('Application Layer'). This layer contains the business logic of the proposed REMONA model which reads in data from a prefabricated data model.

3.4 Data Marts Layer

The application layer of current solutions is as a rule closely connected to the data layer from which it reads in data. For this reason the data model integrated with the application layer so that tasks arising from their integration need not be carried out during implementation of REMONA.

The conceptual model (placed in REMONA data mart) is divided into four areas which are subdivided to the level of the physical data model. The proposed areas are 'Finance and Management Accounting', 'Production Entity', 'Cost Allocation Entity' and 'Other Entity'. The area **'Cost Allocation Entity'** covers a data area related to cost allocations, which include the dimensions 'Cost Centres', 'Cost Drivers'. The area **'Production Entity'** contains data which stem from the core business process of a company. An example is sale of products/services of a company/division and setting their prices and discounts. The area **'Financial and Management Accounting'** covers data related to financial and management accounting.

3.5 Business Concept Model of the REMONA

The application layer (part CPM tools and part REMONA in Figure 1) in which all the logic of REMONA is placed, is based on interconnection of appropriately designed data cubes containing analysed data and parameters for calculations using analysed data and appropriately selected dimensions that enable the data to be analysed.

A key element of the model is appropriate design of the individual cubes.

- Area of historic data which are read in from primary and other systems (in the context of this model usually from tables placed in DWH or data mart),
- Area of Cost Allocation,
- Area of Planning and Modelling, which addresses the task of management profitability.

A Figure 2 shows selected cubes and dimensions of the model REMONA.

3.6 Presentation Layer

We expect in the proposed model the use of two basic types of presentation and analytical cum presentation layers - native tools of the selected environment, such as Cognos Express/TM1, the tool Microsoft Excel.

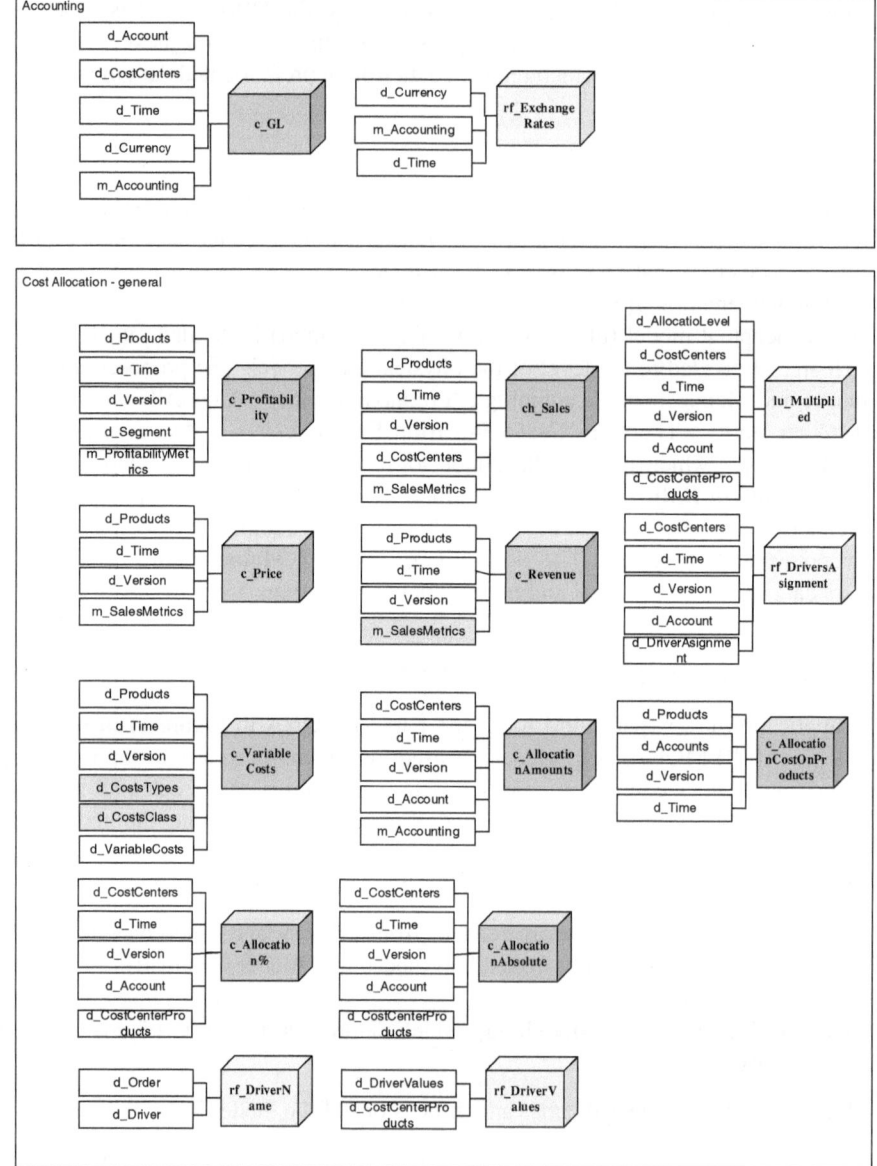

Fig. 2. REMONA – multidimensional cubes and dimensions. Source: Authors

3.7   Model

The current form of the model is represented by the following snapshot of the development/user interface in the tool Cognos Express Performance Modeler. In the left part we see the structure of the multidimensional cubes in which calculations are performed and data are stored. On the right individual parts of the model are developed in individual tags. The active tag represents the structure of one of the selected dimensions.

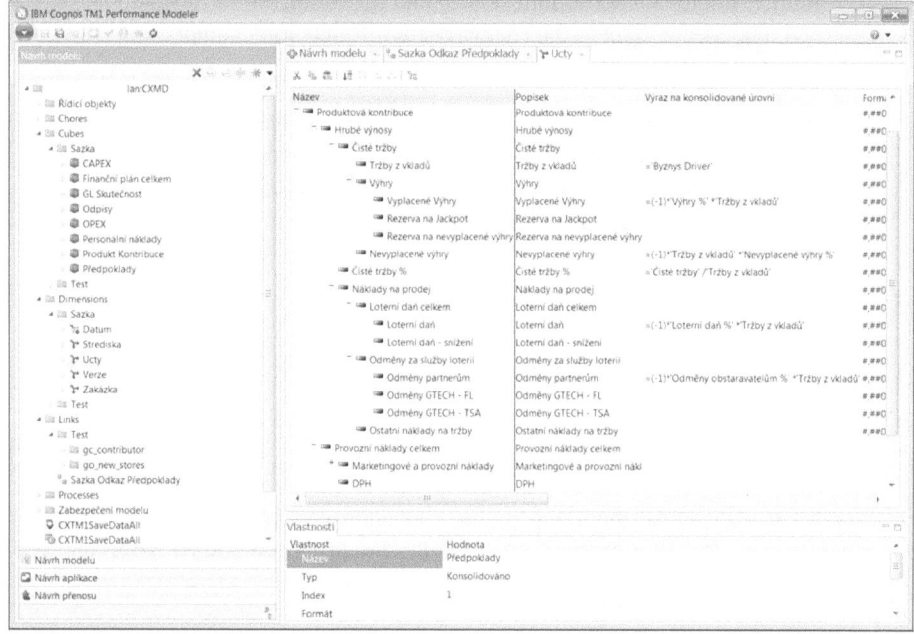

Fig. 3. REMONA – Cognos Express Performance Modeler   Source: Authors

4   Conclusions

Having studied the literature, identified shortcomings and experience from designing other reference models we expect benefits from the proposed model for theory and practice.

The model will serve for practical use in companies and on another level it will be filled with illustrative data available to students who will be able on its basis to more easily visualize the issue of cost allocation and profitability and support for these tasks with technical means. We expect to have for practical use:

- Identification of key metrics which are useful to address the issue of cost allocation, profitability, and related analytical tasks.
- Proposed dimensions necessary to carry out the economic tasks.
- Reference model which will permit:
    - heighten awareness of the relationship between corporate informatics and other organizational units of a company by designing drivers and activities which corporate informatics provide and other parts of the company use,

— put into operation a tool in support of management decision-making in the area of PI.

- Methodology for implementation of the proposed model in a company – determining implementation milestones and possibilities of adaptation of the proposed model according to the characteristics of the target company.

The benefits of the proposed model for theory can be formulated in these aspects:

— Research into cost allocation and profitability in the context of PI.
— Expansion of the theoretical basis of cost allocation and profitability in PI.
— Determining appropriate dimensions and their characteristics.
— Design of the model and its architecture.

5   Directions of Further Research

The further design of the model will consist in answering these questions:

- What are key 'Cost Drivers' enabling cost allocation of PI services?
- What are key activities in the provision of ICT services and creation of PI products?
- Which tasks are most important in cost allocation and profitability management? What is the optimum way of integrating the proposed model with the ICT architecture of a company?
- Pilot implementation of REMONA in selected companies.
- Verification of the model based on experience from pilot implementations and proposition and making of the necessary modifications.

References

1. Dimon, R.: Enterprise Performance Management Done Right: an operating system for your organization. Wiley (2013) ISBN: 978-1118370759
2. Cookins, G.: Performance management: integrating strategy execution, methodologies, risk and analytics. Wiley (2009) ISBN: 978-0470449981
3. Fischer, J., Novotny, O., Doucek, P.: Measurement of Economic Impact of ICT: Findings, Challenges, Perspectives. In: IDIMT 2013 Information Technology Human Values, Innovation and Economy, Prague, September 11-13, pp. 59–62. Trauner Verlag, Linz (2013) ISBN: 978-3-99033-083-8
4. Chandler, A.: Fundamentals of CPM. In: Gartner Amsterdam BI Summit. Gartner, Amsterdam (2007)
5. Chen, Y., Zhu, J.: Measuring Information Technology's Indirect Impact on Firm Performance. Information Technology and Management 5 (2004) ISSN: 1385-951X
6. Muhammad, A.B.: Corporate Performance Management & Impact of Balanced Scorecard System. European Center for Best Practice Management. Research Paper: RP ECBPM/0029 (2010), http://ecbpm.com/files/Performance%20Management/ Corporate%20Performance%20Management%20%26%20Impact%20of%20Bal anced%20Scorecard%20System.pdf [cit. October 16, 2013]

7. Variana, H.R., Farrel, J., Shapiro, C.: The Economics of Information Technology. Cambridge University Press, Cambridge (2004) ISBN: 0 521 60521 0
8. Kral, B.: Manažerské účetnictví. Management Press (2010) ISBN: 8072610627
9. Maryska, M., Novotny, O.: The reference model for managing business informatics economics based on the corporate performance management proposal and implementation. Technology Analysis & Strategic Management, 25(2), 129–146 (2013), doi:10.1080/09537325.2012.759206, ISSN: 0953-7325. WOS:000314348500001
10. Maryska, M., Wagner, J.: Reference model of business informatics economics management. Journal of Business Economics and Management 14(4), 1–17 (2013), doi:10.3846/16111699.2013.789449#.UlgySFPymtY, ISSN 1611-1699
11. Remenyi, D., Bannister, F., Money, A.: The effective measurement and management of ICT cost and benefits. Elsevier (2007) ISBN: 0750683287
12. Lomerson, W.L., Tuten, P.M.: Examining Evaluation Across the IT Value Chain. In: Proceedings of the 2009 Southern Association of Information Systems Conference, Savannah, GA, USA, pp. 124–129 (2009)
13. Strizova, V.: Organizační struktury. In: Doucek, P., Maryska, M., Nedomová, L. (eds.) Informační Management v Informační Společnosti, 1. vyd, p. 264. Professional Publishing, Praha (2013) ISBN: 978-80-7431-097-3
14. Turban, E., Leidner, D., McLean, E., Wetherbe, J.: Information Technology for Management: Transforming Organizations in the Digital Economy. John Wiley & Sons (2007) ISBN: 978-0471725985

A Systematic Approach
for Evaluation and Selection of ERP Systems

Sascha Alpers, Christoph Becker, Esmahan Eryilmaz, and Thomas Schuster

FZI Forschungszentrum Informatik am Karlsruher Institut für Technologie
Haid-und-Neu-Str. 10-14, D-76131 Karlsruhe, Germany
{alpers,cbecker,eryilmaz,schuster}@fzi.de

Abstract. Companies utilize advanced information systems, like Enterprise Resource Planning (ERP) systems, in order to manage their resources and improve their competitiveness. Evaluation, selection and implementation of suitable ERP systems, however, is a challenging task for many companies, especially small and medium sized enterprises (SME). A highly non-transparent ERP-market with a lot of different vendors and system versions, the variety of possible selection criteria as well as individual requirements of the company form a complex decision process. Particularly small and medium companies need support in this phase as it is a critical investment and an important success factor. In this article we will present a systematic approach for evaluation and selection of ERP systems. Our methodology offers decision support for acquisition of a new ERP system by consideration of individual requirements particularly including the company's current processes and IT infrastructure. It is driven by a multi-step process, first conducting a pre-selection of ERP systems, subsequently narrowing down in a second negotiation phase so that finally a suitable ERP system is chosen.

Keywords: Enterprise Resource Planning (ERP), software evaluation, software selection, multistage decision-making, SCAPE, Selection Approach for ERP systems.

1   Introduction

Nearly every company uses business software in order to manage their resources and improve value creation. Selection and implementation of appropriate Enterprise Resource Planning (ERP) systems is a challenging task as it is a critical investment and an important success factor. Especially for small and medium enterprises (SME), which generally possess a smaller amount of resources, costs associated with the acquisition of an ERP system are considered to be important. Thus for SMEs an inadequate ERP system acquisition can put the company's existence at risk [1]. In the most cases it is more efficient to buy and adopt standard software than to invest in own development activities. Even adapting own business processes to the capabilities of a particular standard software system can be more economic than developing own software or investing high efforts into customizing individual activities. Nevertheless,

S. Wrycza (Ed.): SIGSAND/PLAIS EuroSymposium 2014, LNBIP 193, pp. 36–48, 2014.

it is important to sustain existing unique selling propositions or to create new unique features of the products or services that the company wants to place on the market. The decision, which ERP system is the most suitable option, is a complex task and is strongly influenced by the company's individual requirements and business processes. Moreover, the decision affects the level of initial and running cost e.g. maintenance costs [2]. Due to the variety of available selection criteria as well as a highly non-transparent ERP market (comprising more than 100 different ERP systems) a well-defined and structured selection process can be considered a fundamental necessity to make an adequate decision. For this reason we have developed SCAPE – Selection Approach for ERP systems in order to guide small and medium enterprises in the selection process. In addition to this selection method, we built-up a comprehensive database covering systems and vendors of ERP software systems.

The rest of this paper is structured as follows: In section 2 we present our standardized approach for analyzing the needs of SME and provide a general overview on the selection process. In section 3 we describe the critical aspects of the process in more detail. In particular this section focuses on the integration of the company's business processes, requirements and interfaces as well as the application of the actual ERP selection. In section 4 we demonstrate the feasibility of our approach, refer to results of recently conducted projects and apply the approach in a realistic case study. In section 5 we review related work and figure out the main differences to our approach. Section 6 concludes the paper and presents an outlook on future work.

2   A Standardized Approach to Select Suitable ERP Systems

Within various consulting projects we derived SCAPE as standardized methodology to conduct preliminary analysis in order to support the selection of suitable ERP systems for small and medium enterprises. Based on the requirements gathered in the business process and requirements analysis the approach iteratively narrows down the selection of appropriate ERP systems (multi-step process). After individual prioritization of different criteria some systems are chosen for closer examination. Finally three vendors with the highest requirement coverage are invited to present their product. On the basis of available analysis results as well as the product presentation a final recommendation for action is derived. An overview on the approach and comprised steps is presented in Figure 1.

The approach starts with a business process analysis. Based on customer or company internal interviews current business processes are modelled and the business strategy as well as relevant interfaces to existing IT-systems are captured. The business process analysis helps to identify areas of potential improvement and clarify which processes have to be supported by the new ERP system.

During the second step a requirements analysis is conducted based on the previously collected information. Subsequently, functional requirements, environmental constraints as well as required interfaces are gathered in order to derive an individual criteria catalogue. Each criterion is weighted according its priority reflecting the corresponding business strategy.

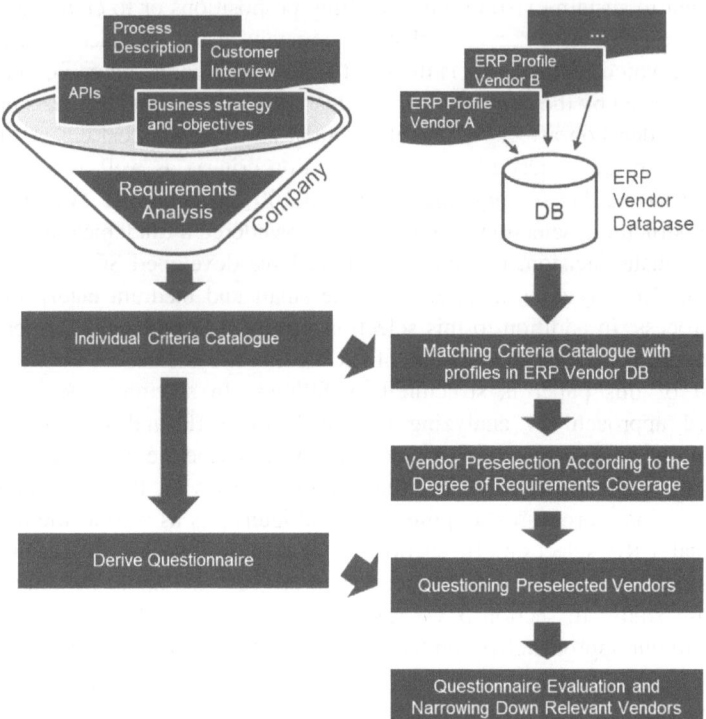

Fig. 1. Standardized approach to select suitable ERP systems

Afterwards, the pre-selection of ERP systems is performed (cf. Figure 1). Therefore the approach makes use of a database for ERP systems and corresponding capability profiles. The company's individual criteria catalog is checked against the database of ERP systems and the degree of requirements coverage is calculated. In addition to the compliance with predefined criteria the approach puts a strong focus on the interfaces that are supported by the ERP system as well as associated license and maintenance costs. ERP systems, which are compliant to the company's must have criteria catalogue, are then added to the list of pre-selected ERP systems.

Based on the criteria catalogue and data collected in previous steps a questionnaire is created and presented to the pre-selected vendors of ERP systems. The questionnaire is an integral part of the call for bids and the results are subject to the contract negotiation. Furthermore, it provides relevant reference to enable constructive discussion in the procurement process and final selection of an ERP system. Each step will be presented in detail in subsequent sections.

3   Assessment Procedure SCAPE

In this section our suggested assessment procedure SCAPE is presented. In preparation for the ERP selection initially the company's current situation is analyzed in the light of current systems and business processes. As highlighted above capturing requirements for ERP systems requires identification, modelling and analysis of relevant business processes. Throughout this analysis pre-existing systems and their support of current and future ERP-related business processes has to be considered in order to prepare a valid choice of an ERP system. Detection of processes and systems is an important step in determining the requirements, and will therefore be described below. In order to systematically survey the analysis the current situation is concluded by grouping the identified requirements into functional and non-functional requirements as well as general conditions and interfaces to existing systems (which have to be integrated according to business processes or replaced by the new system). We believe that it is important that the decision and its main steps are transparent for the company and its stakeholders. So the selected and prioritized criteria must be presented and discussed with the stakeholders. Subsequently a pre-selection of ERP systems is automatically created by combination of collected requirements and an ERP system catalogue (database). Currently we created a catalogue with 61 characteristics of more than 140 ERP systems. This information is partly based on information of ERP system vendors and extended by steady research and evaluation. Maintaining a catalogue of ERP systems is therefore not manageable for companies that seek for appropriate systems, but has to be implemented by research institutes in order ensure independent and unbiased analysis results. Finally the automated pre-selection has to be initially presented to the customer company and can then build the basis of the actual vendor selection, which is driven by concrete offers and interviews before a system is chosen and acquired. It is important that these steps are accompanied by independent experts.

3.1   Identification of Current Business Processes and IT Systems

As part of an initial analysis, the identification of currently utilized systems and pre-existing business processes is a first step in order to derive requirements for an ERP system, which suits the needs of a company. Additionally all identified artefacts of this phase can serve as basis for further discussions and design decisions in later steps.

In order to conduct this first analysis the experts have to examine the company's knowledge base and extend this case based by interviews with the company's domain experts. If the company has correct and fine grained business process models available the analysis of utilized systems can be derived immediately. If such business process models are not available the analysis will start by identification and modelling of business processes. As modeling language most common notations will meet the requirements. We won't cover a detailed language comparison in this article, however, we would suggest usage of either BPMN (Business Process Model and Notation) [3], EPC (Event-Driven Process Chain) [4]or Petri Nets [5]. BPMN is a good candidate to capture not only domain but also technical aspects. Unfortunately, models tend to be too complex to be understood by domain experts, thus the analyst (model expert) has to compensate this. Nowadays BPMN can be considered as becoming a de facto standard. It is also used by many ERP vendors, who allow direct workflow execution based on BPMN models. EPC tend to be more simplistic, thus the analysis of utilized systems and technical interfaces won't be detailed and has to be captured in a second step. Petri Nets are well known in science, their advantage is the possibility of formal analysis (e.g. model checking, soundness or liveness). With our approach the key idea is that all captured artifacts, models and analysis results can be easily re-used in later phases. In the end the final results of this phase are:

- Business process models (of ERP relevant business processes)
- Map of currently utilized systems (including systems which should be replaced, are currently under development or in introduction)
- Overview of interfaces utilized and offered by these systems
- Analysis report of business processes and systems

It is worth to mention that this phase reveals only the current status of the company in terms of business processes and systems as well as potential weaknesses. We consider this analysis as important because design decisions or choice of a new ERP system will affect other systems and employees (how are involved in process execution) as well. Therefore the choice of an inadequate ERP system, that does not fit the company's prerequisites, can create high integration efforts or its introduction can even be refused by key employees.

3.2   Requirements Analysis

The requirement analysis is driven by the company's strategy which determines the importance of all considered relevant criteria. Throughout the requirement analysis three categories of requirements are considered: 1) functional requirements 2) external regulations and 3) interfaces that have to be supported. The analysis of these criteria is not sequential but may be driven concurrently (some criteria may be derived by criteria of other categories). We will detail inquiry of these categories within the following subsections.

3.2.1 Functional Requirements

All functional requirements are classified by categories that are common for ERP systems. These categories will match attributes of our ERP database (see section 3.3). In our approach we are currently covering 61 attributes within the following six categories:

- General Requirements
- Human resources management
- Data processing and management
- Sales / Marketing / CRM
- Controlling
- Industry Sector specific category; sub categories e.g.:
  - Production
  - Services
  - Trading

In dependence on the industry sector in which the company is operating only the specific sub category or categories of *Industry Sector specific category* are considered and evaluated. Each attribute that is evaluated is rated according the following scale:

A $\cong$ highest priority;

B $\cong$ medium priority;

C $\cong$ low priority;

Z $\cong$ additional function currently without priority

Additionally for each attribute we rated if a third party system is involved and if so, which of the systems identified in our previous step is affected. Furthermore we rate if the third party system is leading data management (this means if it is the primary source of data processed according to the rated attribute). The rating has to be examined and evaluated in cooperation with the company that seeks a new ERP system.

3.2.2 External Regulations

In addition to the functional requirements external regulations have to be identified. External regulations can be given by legal restrictions, company compliance rules as well as operations or business strategy. A weighting of these criteria is typically not necessary, since they usually must be met by the new ERP system. All identified criteria are thus re-used in the next selection steps as exclusion criteria. If necessary, however, rating would also be possible. It should be mentioned that fulfillment of these criteria should also be monitored during contract negotiation between company and ERP system vendor. In this case there is no set of criteria which can be considered common for all companies, typical requirements may be:

- Interface & dialogues in the company's main language
- User support in this language
- Service Support Administrators
- Clients fully compatible with a certain environment (e.g. an operation system or a browser engine)
- On-premise installation
- Privacy and protection of employee data
- Data may not leave the company
- No data access for employees of the system vendor
- Redundant system operation and automated backups (according a certain policy)
- The system / the systems maintenance agreement has to be priced below a certain amount

3.2.3 Interfaces

For successful operation of an ERP system, it is also important to know relevant interfaces to external systems. Interfaces to pre-existing systems should be obvious by the previous identification of current business processes and IT systems (as outlined in section 3.1). During this analysis interfaces to systems that are planned to be introduced in the future have to be added, if necessary. All relevant interfaces to third party systems have to be addressed correctly by the ERP system. Similarly to external regulations a common set of interfaces cannot be given, nevertheless there are some typical interfaces in the context of ERP systems, such as interfaces to:

- Human Resource Management Systems
- Document Management
- Office and Groupware Systems
- Wiki Systems
- Telephone System

Of course functions of aforementioned third party systems and associated interfaces may be obsolete, if the ERP system will offer those services. Thus if it is planned to replace a pre-existing system by the new ERP system, this has to be covered by identification of functional requirements (according to section 3.1). All identified interfaces have to be rated according the same scale as introduced in section 3.2.1. Additionally to identification of interfaces their technical implementation has to be taken into account (e.g. web service interface and protocol format) in order to rate if the interface will be addressed properly by the new ERP system. If the interface is not addressed in the standard version of the ERP system, it may still be checked if the system can be extended or modified in order to meet these requirements.

3.3   Pre-selection of ERP Systems

As already described, the outcome of the requirements analysis step is an individual criteria catalogue covering the particular needs and conditions of the company. Individuality is granted by the opportunity of weighting a specific criterion. In the next step this individual criteria catalogue is used to conduct an automated pre-selection of ERP-Systems. By mapping this catalogue to our ERP-database we can create an initial set of valid candidates which have to be evaluated in further steps.

In order to enable an automation of the pre-selection step we built up a comprehensive database which integrates information provided by ERP system vendors and findings of further research and evaluation. Currently this database covers characteristics of more than 140 ERP-Systems and is being steadily extended and updated. An outline of the general structure with the main categories and applied criteria can be seen in Figure 2.

This integrated and structured data set does not only enable a direct and objective comparison between different candidates, features and characteristics, but also an automated pre-selection by applying business intelligence (BI) methods. The individually weighted evaluation criteria within the criteria catalogue are compared with the information stored in the database by applying an appropriate combination of various BI methods. In doing so, we calculate a degree of coverage for each ERP system and an appropriate ranking list.

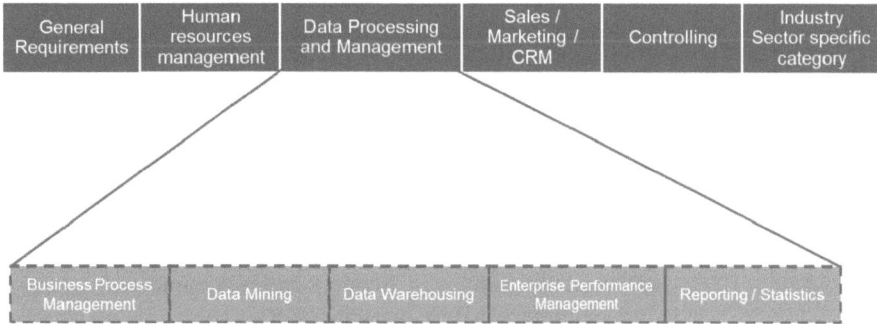

Fig. 2. General structure of the selection criteria set

The size of the pre-selection set is also automatically determined by applying project and company related information. Insofar it has to be underlined, that the pre-selection is based on technical rather than economic aspects. The fixed as well as the running costs become an essential part in the contract negotiation with vendors of the pre-selected ERP systems.

4   Case Study as Evaluation

In this section we will demonstrate how our method is applied in a brief case study. We won't illustrate how to gather the requirements (this was highlighted in section 3 already) but present specific requirements that will serve as a reduced and company individual example. All results presented below are taken from a project conducted for a SME located in Germany. Since we do not intend to advertise products, we anonymized the company's name (SME) as well as the selected ERP vendors and system versions. In this case study the company is a service company which offers services only to other business partners (companies, B2B). In particular the services include standardized data processing, data analytics and consulting for a limited number of customers. Therefore the subcategory *Services* of the industry sector categories is selected. For the sake of simplicity we will outline only some attributes and associated priorities to highlight how weighted attributes can be utilized to represent the company's strategy (see Table 1). In the example, CFO is an example for an attribute which is not relevant for the company. It would not be analyzed further. The attributes correspond with the 61 characteristics stored in the vendor database. Hence all attributes serve as criteria which are compared to the systems in our vendor database. Besides the attributes outlined below, we captured 30 attributes to calculate the system pre-selection properly.

Table 1. Attributes rated according to company strategy

Category	Functionality	Description	Rating
Industry sector, subcategory: *Services*	Order Allocation (OAL)	Allocate Orders to Roles	A
Sales / marketing / CRM	Master Contract Management (MCM)	Administration of framework agreements	A
Sales / marketing / CRM	Management of subscriptions and frequent orders (MSF)	Planning subscriptions and frequent orders (reminders, special offers, …)	B
Sales / marketing / CRM	Capacity Forecasting (CFO)	Analyzing Trends for occupancy rate of staff and other resources	--
Sales / marketing / CRM	Customer Data Management (CDM)	Management of core data of customers.	Z
human resources management	Personnel Assignment Planning (PAP)	Managing assignment of tasks to staff.	A
General Requirements	Ability to Customize ERP system (ACE)	Customization according current and future business processes by extension module or third party system	B
Controlling	Pre calculation (PRC)	Calculate the effort of tasks before they done.	C
Controlling	Post control (POC)	Calculate the effort of tasks after they done.	C

If a system fulfills a criterion it is credited with one criteria point (cp). Weighted criteria points (wcp) are calculate by multiplication of each cp with the criterion's corresponding priority (priority A times 7; B times 4; C times 1; Z times 0 because it is currently not in focus). Finally for each system the sum of cp and wcp is calculated. By normalization to a scale from 0 to 1 we create the so called suitability index (si). With the suitability index we enable a comparison of system qualification across ERP acquisition projects (even at different companies). Thus we are able to create a general index of ERP systems. Table 2 outlines the evaluation results of all attributes presented above (Table 1). In Table 3 the aggregated results for all considered systems and all attributes captured in this project are presented.

Table 2. Evaluation of five case study systems

Criteria	Priority	System 1		System 2		System 3		System 4		System 5	
OAL	A (7)	1	7	1	7	0	0	0	0	0	0
MCM	A (7)	1	7	1	7	1	7	0	0	0	0
MSF	B (4)	0	0	0	0	1	4	0	0	1	4
CDM	Z (0)	1	0	1	0	0	0	0	0	1	0
PAP	A (7)	1	7	1	7	1	7	0	0	0	0
ACE	B (4)	0	0	1	4	1	4	1	4	1	4
PRC	C (1)	0	1	1	1	0	0	1	1	1	1
POC	C (1)	0	1	1	1	0	0	1	1	1	1
cp / wcp	max. 8 / 31	4	21	7	27	4	22	3	6	5	10
Si	[0;1]	-	0.68	-	0.87	-	0.71	-	0.19	-	0.32

Table 3. TOP 5 systems after analyzing all selected[1] criteria

Rank	System	cp (max 30)	wcp (max. 128)	si [0;1]
1	System 2	22	105	0.82
2	System 12	24	98	0.77
3	System 3	17	95	0.74
4	System 5	18	81	0.63
5	System 54	15	72	0.56

It is obvious that the decision process is not properly supported if the company's strategy is left out. In Table 3 for instance System 12 would be rated higher than System 2, which actually would reflect the company's requirements better than System 12 (see wcp). With weighted criteria points the most suitable systems can be pre-selected in respect of strategy and most important requirements. Typically we suggest choosing up to ten systems to be further evaluated by questionnaire in order to keep the effort for analyzing the feedback within a limit. The questionnaires can be used to pinpoint fine-grained company requirements which cannot be derived from database analysis. Typical examples are the support of specific interfaces, for instance an interface to export accounting information to a third party system (e.g. of a tax advisor) using a specific format. Based on the questionnaire feedback the systems are reevaluated. Only a few (we suggest two or three) vendors should then be invited to present their system (negotiation phase).

5   Related Work

Several approaches and best practices to support software selection in general and in particular the selection of ERP systems exist in literature. Current research activities

---

[1] In this case 30 of 61 criteria are identified as relevant for the company.

as well as a number of different studies analyze the selection of ERP systems for various countries or specific industrial sectors [6]–[13].

A detailed review of these approaches reveals that there is no holistic methodology supporting a company in the selection of an ERP system in respect to a given strategy and in correspondence to business processes. In particular there is no satisfying approach, which analyzes the selection process from an enterprise perspective [13]. The majority of investigated approaches are based upon a set of selection criteria which are used to calculate a score and rank the considered ERP system candidates. The selection criteria are generally derived from literature and organized according to various categories. A common differentiation is based on ERP vendor and ERP product criteria [9], [10]. However, categorization according to functionality, technical architecture, costs, service and support conditions or usability can be found as well.

The evaluation of selection criteria has to deal with various difficulties. Some criteria are not easy to quantify, others are ambiguously formulated and lead to trouble in the evaluation process. For example a selection criterion regarding the friendliness of the user interface is widely subjective and based on opinions or impressions of the evaluating person. For an objective evaluation and choice the criteria should be based on facts defined with a more formal approach. Additionally the majority of currently available approaches lack an easy way to customize or expand the set of selection criteria. However, this is an essential requirement as the relevance of each criterion strongly depends on the company's individual characteristics and conditions. Thus it is necessary to link each criterion with a relative weight whose value can be specified in a pre-step of the selection process according to specific needs and resources of the company.

A further shortcoming of current approaches concerns the pre-selection of an initial set of valid ERP candidates. Most approaches select a number of vendors in a pre-selection step, which is not described in detail, and continue their selection process with these candidates. Even formalized approaches, using various multi-attribute decision-making methods like [10], [11], [14]–[16] do not reveal their pre-selection steps. This can also be related to the lack of available, comprehensive datasets of ERP vendors enabling a direct, objective comparison between different candidates, features and characteristics according to respective selection criteria. Without a well-defined procedure the pre-selection of candidates is not transparent to the customer company and in particular can lead to insufficient decisions.

Another important weakness we identified is that there is no sufficient process-oriented approach for the selection of ERP systems, although the business processes of a company represent the most important source for requirements related to ERP-systems. The approach presented in this paper solves these aforementioned issues by explicitly focusing on the pre-selection of ERP systems incorporating a detailed process-driven requirements analysis as well as taking into account a solid reference database of available ERP systems.

6   Conclusion

Throughout this article we introduced a methodology for sound selection of ERP systems. This includes that company specific requirements are captured and utilized to measure conformance of ERP systems. The presented approach is also taking into account the company's current processes and IT infrastructure. Conformance measurement is driven by weighted requirements according to the company's strategy. Finally, this results in decision support for acquisition of a concrete ERP system. The decision support is driven by a two-step process; at first a vendor shortlist (pre-selection) is created which is narrowed down in a second negotiation phase that is concluded by choice of a system. Currently the weighted requirements used in our approach have to be determined manually, this causes additional effort. Therefore, as next steps we plan to elaborate on an enhanced and formalized method to map company strategy automatically to weighted requirements. So far we applied our methodology for several companies; however, we plan to benchmark our results with other approaches. Therefore we intend to examine which methodologies are being used to choose ERP systems and how often they are applied successfully.

References

[1] Hustad, E., Olsen, D.H.: ERP Implementation in an SME: A Failure Case. In: Devos, J., van Landeghem, H., Deschoolmeester, D. (eds.) Information Systems for Small and Medium-sized Enterprises, pp. 213–228. Springer, Heidelberg (2014)

[2] Hendricks, K.B., Singhal, V.R., Stratman, J.K.: The impact of enterprise systems on corporate performance: A study of ERP, SCM, and CRM system implementations. J. Oper. Manag. 25(1), 65–82 (2007)

[3] OMG, Business Process Model and Notation (BPMN), Version 2.0.2 (2013)

[4] Keller, G., Nüttgens, M., Scheer, A.-W.: Semantische Prozessmodellierung auf der Grundlage "ereignisgesteuerter Prozessketten (EPK)". Iwi, Saarbrücken (1992)

[5] Reisig, W.: Understanding Petri Nets: Modeling Techniques, Analysis Methods, Case Studies. Springer, New York (2013)

[6] Hecht, B.: Choose the right ERP software-ERP vendor selection can be filled with vendor hype, internal political agendas, and unmet expectations. Today, as ERP has risen in strategic significance, choosing. Datamat.-Highl. Ranch. 43(3), 56–61 (1997)

[7] Tsai, W.-H., Lee, P.-L., Chen, S.-P., Hsu, W.: A study of the selection criteria for enterprise resource planning systems. Int. J. Bus. Syst. Res. 3(4), 456–480 (2009)

[8] Ratkevičius, D., Ratkevičius, Č., Skyrius, R.: ERP Selection Criteria: Theoretical and Practical Views. Ekonomika/Economics 91(2) (2012)

[9] Poon, P.-L., Yu, Y.T.: Procurement of Enterprise Resource Planning Systems: Experiences with Some Hong Kong Companies. In: Proceedings of the 28th International Conference on Software Engineering, New York, NY, USA, pp. 561–568 (2006)

[10] Silva, J.P., Goncalves, J.J., Fernandes, J.A., Cunha, M.M.: Criteria for ERP selection using an AHP approach. In: 2013 8th Iberian Conference on Information Systems and Technologies (CISTI), pp. 1–6 (2013)

[11] Shuai, J.J., Kao, C.Y.: Building an effective ERP selection system for the technology industry. In: IEEE International Conference on Industrial Engineering and Engineering Management, IEEM 2008, pp. 989–993 (2008)

[12] Holland, C.P., Light, B.: A critical success factors model for ERP implementation. IEEE Softw. 16(3), 30–36 (1999)
[13] Lech, P.: Information gathering during Enterprise System selection - insight from practice. Ind. Manag. Data Syst. 112(6), 964–981 (2012)
[14] Wang, T.C., Chiang, Y.C., Hsu, S.C.: Applying incomplete linguistic preference relations to a selection of ERP system suppliers. In: 2007 IEEE International Conference on Industrial Engineering and Engineering Management, pp. 119–123 (2007)
[15] Rouyendegh, B.D., Erkan, T.E.: ERP System Selection by AHP Method: Case Study from Turkey. Int. J. Bus. Manag. Stud. 3(1) (2011)
[16] Wei, C.-C., Chien, C.-F., Wang, M.-J.J.: An AHP-based approach to ERP system selection. Int. J. Prod. Econ. 96(1), 47–62 (2005)

Job Satisfaction and Ethical Behaviors Premises of IT Users Insight from Poland

Alicja Keplinger, Jolanta Kowal, Emilia Frątczak,
Karolina Ławecka, and Paulina Stokłosa

University of Wroclaw, Department of Historical and Pedagogical Sciences,
Institute of Psychology, Wroclaw, Poland
a.keplinger@psychologia.uni.wroc.pl, jolakowal@gmail.com

Abstract. The paper concerns ethical, human and organizational aspects of IS development. The aim of the research is to examine the dependency of job satisfaction (JS) and organizational citizenship behaviors (OCB) with its two dimensions (individual and organizational), among IT users in Poland, a transition economy. The research results complement the gap in the scientific literature, that concerns the psychosocial characteristics of IT users, as ethical attitudes and job satisfaction components. The authors elaborated questionnaire of JS including such dimensions like relations between employees, supervisors' managing style, the organization of work in the company, information and communication within the company, representing the interests of the employee, creating opportunities for staff development (motivation, evaluation, promotion) and subjective JS. The analysis is based on a survey conducted among 362 IT users in south-west region of Poland. The results of this survey show that job satisfaction has an impact on ethical behavior premises of IT users in transition economies.

Keywords: ethical behavior premises, IT, IS Development, job satisfaction, organizational citizenship behavior, Poland, transition economy.

1 Introduction

The information system (IS) may be defined as a social human activity system created by elements which belong to five classes, including: data, methods, information technology, organization and people [22, 27, 31]. The information technology (IT) comprises computer hardware and software used to create, transmit, display and secure information, combining telecommunications, utilities, and other technology-related information [11,29].

Information systems are characterized by continuous development [10,11] that is related to the issues of technology and organizational factors, such as human capital. The organization development is often limited by the technology and infrastructure, knowledge or competence. Information systems dynamically evolve towards the business environment in relation to market requirements [33]. That's why it is important to monitor IT users skills, their organizational citizenship behaviors (OCB) and job satisfaction (JS).

S. Wrycza (Ed.): SIGSAND/PLAIS EuroSymposium 2014, LNBIP 193, pp. 49–64, 2014.
© Springer International Publishing Switzerland 2014

The authors of the paper are interested in the development of information systems (IS) *as a creative effort that comprises the expertise, insights, and skills of employees concerned with the need of improving for business* [33]. The authors are especially interested in IT users' development in the sphere of JS and OCB. In need of current research the authors distinguish two groups: using IT intensively called "IT users" (working a minimum of 20 hours per week) and extensively (less than 20 hours per week) signed as "Others". However "IT users" in current study are not understood as IT professionals. IT users can skillfully use hardware and software (created by the IT professionals), but they do not create systems and programs themselves.

The authors assume: 1) the impact of JS on OCB, as also suppose that IT users working more intensively in IS significantly differ from all other employee groups, what may be concerned with their higher competences and expectations; 2) IT employee satisfied with the work manifest OCB [16, 18, 19, 20].

The next part of the paper has the following structure: In the proceeding section the authors make a brief review of literature related to JS and OCB. Hypotheses verification is based on the data from a structured survey conducted among 362 IT users employed in enterprises located in south-west regions in Poland. Further, the results of this research are depicted and discussed, with proposition of some ideas for future studies.

2   Literature Review and Hypothesis

The aim of the research is to describe the relationship and examine the dependency between JS and OCB among IT users in Poland, a transition economy. The research results complement the gap in the scientific literature, that concerns the psychosocial characteristics, including ethical attitudes and job satisfaction components of IT users professional group. In particular, the authors were interested in finding the answers to the questions correlated with formulated hypothesis: if job satisfaction (JS) has significant effect on the scale of organizational citizenship behaviors (OCB), and its two dimensions: OCB individual (OCB-I) and OCB organizational (OCB-O) of IT users.

This research includes new elements of OCB sphere: translation, cultural adaptation and applying the Questionnaire of Organizational Citizenship Behaviors (by Konovsky and Organ, [7]), first time in Poland. In addition, the novelty is the assumption that job satisfaction has an impact on OCB (in contrast to other authors proposing opposite dependency) among IT users and de fact that employee and employer describe each other. This method is more objective and reduces personal idealization. In the world literature there are few examples of this approach (for instance [16]) and lack of research on this theoretical model in Poland. Despite of that there is not too much studies on OCB in Polish companies.

Figure 1 shows the theoretical model of JS and OCB relation that has been proven empirically.

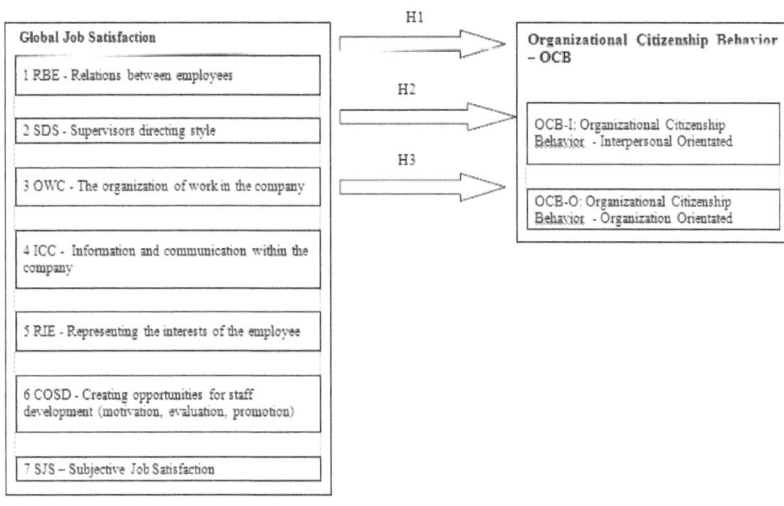

Fig. 1. The model and hypothesis

2.1 Job Satisfaction

The job satisfaction (JS) has been examined by researchers interested in task performance [3, 4, 24, 25], little is known about how certain dimensions of JS influence OCB. Therefore, this study aims to understand the nature of the relationship between JS and OCB within the context - Polish transition economy.

Locke [14] is one of the scholars who defined JS by mentioning that; JS is the pleasant feeling that results from the appraisal of the job or by the job facilities, whereas job dissatisfaction is the unpleasant feeling that results from the appraisal of job such as, frustration or blocking the achievement of values. In our study variable of the JS base on the psychological climate concept by Rosenstiel and Boegel [21]. Authors maintained: organizational climate is a notion referring to the features of the entire internal environment of an organization. as perceived and evaluated by groups of its members. The following most relevant dimensions of organizational climate assessment were emphasized:

1. the relationships among the employees
2. the style of management
3. the work organization
4. information flow and communication within the company
5. the representation of the employees' interests
6. occupational development possibilities for the employees (motivation, evaluation, promotion)

The authors are interested in how IT users and other employees describe level of JS with six dimensions which were pinpointed by Rosentiel and Boegel. The authors claim that organizational climate can be assessed on these six dimensions, and their positive evaluation means JS in the above six aspects.

2.2   Ethical Behavior – OCB: Organizational Citizenship Behavior

Organizational Citizenship Behavior (OCB) has become very popular in psychology and management in developed economies [1, 12, 18, 19, 23, 26, 32]. Organ [17] stated that *Individual behavior that is not explicitly or indirectly recognized by the formal reward system and that behavior play a vital role in the effective functioning of the organization.* He termed this behavior as discretionary behavior. By discretionary, we mean that the behavior which is not part of a formal contract or having proposed set of tasks or activities is rather a discretionary choice of an individual to endorse [18]. Podsakoff [20] defined OCB as behaviors, which are not covered by the formal job description, but usually facilitate the realization of tasks and support in enhancement of these behaviors in organizational settings. This behavior is voluntary and helpful and it is not required by the individuals role or job description. Employees who display OCB can contribute to improving organizational efficiency and effectiveness [15].

Therefore, we hypothesize:

Null Hypothesis H01: Job Satisfaction has significant effect on the OCB of IT users.

Alternative Hypothesis HA1: Job Satisfaction no significant effect on the OCB of IT users.

2.2.1   OCB-I: Organizational Citizenship Behavior - Individual

Some researchers have suggested that OCB fits into two categories. Williams and Anderson [32], divided OCB into two types: (1) behaviors directed at specific individuals in the organization, such as courtesy and altruism (OCB-I); and (2) behaviors concerned with benefiting the organization as a whole, such as conscientiousness, sportsmanship and civic virtue Organizational Citizenship Behavior -Organizational (OCB-O). The authors would like to discuss two types of the behaviors revealed by the Polish employees (e.g., OCBs targeted at individuals (OCB-I) vs. OCBs targeted at organizations (OCB-O). OCB-I helps to maintain a balance in the organization, fostering employee transactions. Proponents of this approach assume that behaviors fall into one of these two categories and that are two higher order dimensions of OCBs likely have different genesis. For example, OCB-I reflecting helping, cooperative behaviors are enacted to benefit other people in some way. OCB-I refers to the behaviors that immediately benefit specific individuals within an organization and, thereby, contribute indirectly to organizational effectiveness [18, 32]. Podsakoff [20] labelled this dimension as helping behavior and defined it as voluntarily helping others with work-related problems.

2.2.2   OCB-O: Organizational Citizenship Behaviors - Organizational

The second dimension of OCB includes behaviors benefiting the organization without actions aimed specifically toward any organizational member or members (e.g., adhering to informal rules, volunteering for company). Podsakoff [20] labelled this organizational compliance as it involves an internalization of a company's rules and policies. Furthermore, Williams and Anderson [32] defined it as behaviors that benefit the organization in general. These behaviors include giving prior notice regarding an absence from work or informally adhering to rules designed to maintain order. Behaviors like conscientiousness, sportsmanship, compliance, and civic virtue are enacted to benefit the larger organization, not specific only for people - are good predictors of OCB-O.

2.3   Job Satisfaction and Organizational Citizenship Behaviors

Empirical studies carried out by various researchers to establish the relationship between OCB and JS [1, 10, 13,17, 20, 25, 28]. The researchers found a strong and positive relationship between OCB and contextual JS [16]. The present study investigates the relationship between two dimensions: OCB-I and OCB-O as the dependent variable and the six facets of JS as the independent variable. Furthermore, in this study, it has been assumed that in the Polish small companies, employees who report high levels of intrinsic satisfaction with their work will be more involved in activities such as helping others who have been absent and defending the organization when others criticize it. On the basis of these assumptions, the following hypotheses have been developed:

Therefore, we hypothesize:

Null Hypothesis H02: Job Satisfaction has significant effect on the OCB-I of IT users.

Alternative Hypothesis HA2: Job Satisfaction has no significant effect on the OCB-I of IT users.

Null Hypothesis H03: Job Satisfaction has significant effect on the OCB-O of IT users.

Alternative Hypothesis HA3: JS has no significant effect on the OCB-O of IT users.

In the present study, we begin to address this gap by conceptualizing an attitudinal mechanism – Job Satisfaction - that is likely to mediate the effect on individual - and OCB-I and OCB-O.

3   Methodology

To conduct the research and verify the three research hypotheses the authors used qualitative and quantitative methods [8] like the method of competent judges and structured survey. The first one was applied in the process of adapting 'OCB' questionnaire and elaborating JS questionnaire. Structured survey was applied in the main study.

3.1   Research Questionnaires

To collect the data for analysis of OCB, the authors applied existing questionnaire - OCB [7] and adapted it to Polish cultural conditions [10]. The authors translated the questionnaire from English to Polish and made cultural adaptation. This process comprised the method of competent judges, items discriminatory power, scales validity (CFA) and reliability (Cronbach's α) analysis. To measure the discriminant validity of the construct the authors apply the Average Variance Extracted (AVE) method, in order to examine whether the amount of variance explained by the construct in relation to the amount of variance due to the measurement error is significant [8, 9, 10]. The AVE results for OCB dimensions were statistically significant and respectively equal to: AVE (OCB) > 0.75, which are quite acceptable results for both dimensions of the questionnaire. The Cronbach's alpha coefficient of the scale for this study was greater than 0.94, the average correlation between items was about 0.33, RMSEA was less than 0.07. CHI^2/DF=2.68<5 [see 9, 10].

Table 1. Items for OCB Questionnaire (Adapted from Konovski and Organ [6, 7] by Keplinger et al. 2014 - unpublished)

Dimension	Variable Name	Items
OCB-I	OCB1	1. Helps others who have heavy work loads.
	OCB2	2. Helps others who have been absent.
	OCB5	5. Helps make other workers productive.
	OCB6	6. Helps orient new people even though it is not required.
	OCB7	7. Shares personal property with others if necessary to help them with their work.
	OCB12	12. Respects the rights and privileges of others.
	OCB13	13. Tries to avoid creating problems for others.
	OCB14	14. Considers the effects of his/her actions on coworkers.
	OCB15	15. Consults with me or other people who might be affected by his/her actions or decisions.
	OCB16	16. Informs me before taking any important actions.
	OCB23	23. Thinks only about his/her work problems. not others.
OCB-O	OCB3	3. Looks for other work to do when finished with assigned work.
	OCB4	4. Always does more than he/she is required to do.
	OCB8	8. Tries to make the best of the situation. even when there are problems.
	OCB9	9. Does not complain about work assignments.
	OCB 10	10. Is able to tolerate occasional inconvenience when they arise.
	OCB 11	11. Demonstrates concern about the image of the company.
	OCB 17	17. Never abuses his/her rights and privileges.

Table 1. (*continued*)

Dimension	Variable Name	Items
	OCB 18	18. Always follows the rules of the company and the department.
	OCB 19	19. Always treats company property with care.
	OCB20*	20. Complains a lot about trivial matters.*
	OCB21*	21. Always finds fault with what the organization is doing.*
	OCB22*	22. Expresses resentment with any changes introduced by management. *
	OCB24*	24. Pays no attention to announcements. messages. or printed material that provide information about the company.*
	OCB25	25. Is always on time.
	OCB26	26. Attendance at work is above average.
	OCB27	27. Gives advance notice when unable to come to work.
	OCB28	28. Maintains a clean workplace.
	OCB29	29. Always completes his/her work on time.
	OCB30	30. Stays informed about developments in the company.
	OCB31	31. Attends and participates in meetings regarding the company.
	OCB32	32. Offers suggestions for ways to improve operations.

All items are measured on a 5-point scale:

For OCB strongly disagree (1). disagree (2). neutral (3). agree (4). strongly agree (5). * Reverse scale items.

To value translation quality, the QOCB was then rendered back in English by a different translator with satisfying result [8,10]. In the present study, two classes of employee behavior were measured. The performance of OCBs covered both OCB-Is and OCB-Os. A 5-point Likert scale ranging from "strongly disagree" to "strongly agree" was also used by respondents when they completed the questionnaire. Higher scores reflected higher levels of OCB. The questionnaire dimensions and items are presented in Table 1.

The second questionnaire (new tool) concerning JS was prepared by authors on the basis of the theory of Rosentiel and Boegel [21]. The AVE results for QJS dimension were statistically significant and respectively equal to: AVE (OCB) > 0.77, the Cronbach's alpha coefficient was greater than 0.92, the average correlation between items was about 0.4, RMSEA was less than 0.09. $CHI^2/DF=3.01<5$ [see 9, 10]. Both the OCB and some JS dimensions of the Questionnaire of Organizational Climate [2] Questionnaires have been widely used in earlier studies concerning different professional groups and cultures and that is the reason of applying them.

Table 2. Items for Job Satisfaction (Elaborated by Keplinger et al. , unpublished, on the basis of the theory Rosentiel and Boegel [21])

Dimension	Variable Code	Questions
1 RBE	JS1 Job Satisfaction	1. Are you satisfied with employee relations?
Relations between employees	JS2	2. Are you satisfied with the atmosphere at work?
2 SDS Supervisors directing style	JS3	3. Are you satisfied with the leadership style practiced superiors?
3 OWC The organization of work in the company	JS4	4. Are you satisfied with the physical conditions of work (eg. Lighting, Temperature, equipment)?
	JS5	5. Are you satisfied with the organization of work?
	JS6	6. Are you satisfied with the nature of the work?
4 ICC Information and communication within the company	JS7	7. Are you satisfied with the progress of information and communication within the company?
5 RIE Representing the interests of the employee	JS8	8. Are you satisfied with the way represent the interests of the employee?
	JS9	9. Are you satisfied with the security occupied the job?
	JS10	10. Are you satisfied with the forms of employment
6 COSD Creating opportu nities for staff developmen t (motivation, evaluation, promotion)	JS11	11. Are you satisfied with your pay?
	JS12	12. Are you satisfied with offered employees opportunities for development (motivation, evaluation, promotion)?
7 SJS Subjective Job Satisfaction	JS13	Rate your satisfaction with the work on a scale of 1 to 5
All items are measured on a 5-point scale: For OCB strongly disagree (1), disagree (2), neutral (3), agree (4), strongly agree (5)		

Table 3. Sample Characteristics

Variables	IT Users		Others	
	Quantity	**Percent**	**Quantity**	**Percent**
Age in years				
less than 20	2	2.4	1	0.36
20-29	36	42.9	50	17.99
30-39	28	33.3	45	16.19
40-49	16	19.1	28	10.07
50-69	1	1.19	29	10.43
Missing data			125	44.96
Gender				
Male	27	32	47	29.75
Female	57	68	110	69.62
Education				
Secondary	11	13.1	50	17.99
Vocational	3	3.57	13	4.68
Technical	11	13.1	9	3.24
Higher Engineering or Bachelor	21	25	31	11.15
Master	38	45.2	86	30.94
Missing data	38	45.2	89	32.01
Position within company *				
Employer	25	30	174	62.59
Employee	59	70	104	37.41

3.2   Research Questionnaire Adaptation and Testing

The source QOCB and QJS were translated from English to Polish using the Translation, Review, Adjudication, Pre-testing and Documentation (TRAPD) team translation process [5, 10]. In 2014, for the goal of adapting the research tools to cultural and market conditions in Poland, the authors conducted qualitative and quantitative initial and pilot studies with the group of competent judges, that had knowledge and practical experience with the economy and social community in Poland [10]. The adaptation comprised some modifications to the target questionnaires, concerning semantic and grammatical aspects of individual items, that better suit to the population of IT users and other employees working in Poland. Respondents working in various companies in Poland were invited to participate in a pilot survey that suggested good results, concerning discriminatory power (AVE), scales validity (CFA) and reliability (Cronbach's α) of the target survey instruments. For all dimensions, standardized Cronbach's alpha coefficients for QOCB and for QJS were greater than 0.7.

3.3   Participants and Data Collection

The sample construction joined the methods of random interpersonal network and sequence sampling with the passive optimal experiment design [8]. The trial of 362 employees was collected via individual direct survey in small and medium-sized companies (SMEs).

Table 4. Descriptive Statistics ($N_{IT}=84$, $N_{OTHERS}=278$)

Codes of Variable	Mean		Median		Standard deviation	
	IT	OTH.	IT	OTH.	IT	OTH.
OCB_I	3.84	3.88	3.91	3.95	0.72	0.64
OCB_O	3.82	3.87	3.79	3.95	0.62	0.54
OCB	3.82	3.88	3.81	3.95	0.63	0.54
RBE*	3.86	3.54	4.00	3.50	0.77	0.86
SDS	3.96	3.89	4.00	4.00	0.99	0.93
OWC	4.03	4.00	4.00	4.00	0.64	0.78
ICC	3.82	3.83	4.00	4.00	0.87	0.97
RIE	3.71	3.76	3.67	4.00	0.82	0.82
COSD*	3.46	2.04	3.50	2.50	0.93	1.93
SO	3.84	3.82	4.00	4.00	0.69	0.88
SUM_SAT	3.52	3.48	3.54	3.54	0.58	0.65

- The differences are marked, on the significance level $p<0.05$

3.4   Statistical Methods. Analysis Results

In the current study for all questions measuring OCB and the JS the variables from the 5-point Likert scale were applied as follows: 1 means "I strongly disagree," and 5 "I strongly agree." The statistical methods were chosen appropriately to measuring variables' scales. They comprised the descriptive statistics, the point estimation, the section estimation and the statistical hypotheses verification.

3.4.1 Job Satisfaction

Due to the authors' results presented in Table 4 and Table 5. IT users in Poland are mostly satisfied in each dimension of JS (68%) in comparison to other employees (39%). It is rather similar to the results reported from the developed countries [4]. Subjective JS was also high (IT users 73,8 %, others 42,8%). The average for global JS in the group of IT users was equal to 3.5 (in the scale from 1 to 5), so greater than middle point. IT users are especially more satisfied with employee relations (mean for IT users m=3.41 versus others m=3.12) and the atmosphere at work (IT users m=4.31 and others m=3.95). They are less satisfied with the pay (the IT users m=3.49 and others m=3.69). The IT users were most satisfied with the organization of work in the company (The organization of work in the company - OWC, IT users 77,4 %, others 44%), a little less but comparatively with supervisors managing style (supervisors directing style - SDS, IT users 75,0 %, others 41,4%), and then with relations between employees (relations between employees - RBE, IT users 71,4 %, others 32,6%), with information and communication within the company (Information and communication within the company - ICC, IT users 66,7 %, others 41,7%), with representing the interests of the employee (IT users 65,1 %, others 36,5%). IT users were much less satisfied with creating opportunities for staff development motivation, evaluation, promotion (creating opportunities for staff development (motivation, evaluation, promotion) - COSD, IT users 50,0 %, others 36,5%).

The means were respectively equal for RBE: relations between employees - IT users m=3.86; others m=3.54, p<0,05.The results for COSD: creating opportunities for staff development (motivation, evaluation, promotion) – were for IT users m=3.46; others m=2.04, p<0,05. The analysis of the results suggests that IT users are satisfied with relations between employees (71%) that is comparable (but with higher percent) to Kowal and Roztocki [10] studies concerning IT professionals (49%). The mean (3.86) and median (4.00) of current analysis are higher than the non-opinion level of 3.0.

The authors believe that the above results indicate that the higher the level of competence of IT users, in comparison to the others, may create greater possibilities of development and support for building the relationships between the employee. This conclusion was also confirmed by observed positive correlation between satisfaction with employee relations and the level of competence of IT users concerning capability of using information technology, position and seniority (multiple R-square was equal to 0.3, p<0.05).

3.4.2 Organizational Citizenship Behavior

In the case of OCB the authors observed that IT users in Poland evaluated their behaviors rather positively (68%). This is indicated by the median of 3.81 (on the scale from 1 to 5), which is close to the mean of 3,82. Both the median and mean over the non-opinion level of 3.0 and skewness equal to -0.35 show the domination of positive results over the average. There were only 11.7% negatively evaluated opinions about OCB.

3.4.2.1 Citizenship Behavior – Organizational. Analysis of OCB-O results suggests that IT users (similarly as others) in Poland evaluated their behaviors rather positively (68.14%). This is shown by the median of 3.79 similar to the mean of 3.82, higher than the non-opinion level (3.0). Skewness equal to -0.35 indicates the domination of positive results. The negatively evaluated opinions about OCB-O were approximately equal to 11.96%. Thus, in this case IT users perceived the organization behaviors as definitely positive. The authors observed that IT users have lower average in the range of OCB-O (m = 3.82) than others (m = 3.91, p <0.05). Persons other than IT users rate higher and more optimistic the level of OCB for co-workers. The results are comparable to research of Roztocki and Kowal [10] on the ethical level of optimism among IT professionals that are rather pessimistic.

3.4.2.2 Citizenship Behavior – Individual. Analysis of the results of citizenship behavior individual showed rather positive attitudes IT users and other employees (means: for IT users m=3.84, others m=3.88). Skewness equal to -0.37 indicates the domination of positive results. The negatively evaluated opinions about OCB-I were approximately equal to 18.1%. It means that IT users perceived the good behaviors and assessed themselves and other employees in positive way.

3.4.3 The Effect of Job Satisfaction on Organizational Citizenship Behavior

To verify the three hypotheses and answer the research questions, the authors applied Pearson's linear correlation coefficient and results were shown in Table 5. The authors interpreted the component dimensions, if the correlations were significant and strong enough.

Table 5. Correlation Matrix (Pearson Correlation Coefficient. N=84)

	OCB-I	OCB-O	OCB
SO	0.61	0.53	0.58
RBE	0.20	0.20	0.20
SDS	0.45	0.41	0.44
OWC	0.43	0.41	0.43
ICC	0.56	0.54	0.57
RIE	0.71	0.59	0.66
COSD	0.57	0.53	0.57
Global JS	0.63	0.56	0.61

OCBs were positively correlated with all dimensions of JS with a Pearson correlation coefficient of 0.61 for global JS and with other dimensions like representing the interests of the employee (0.66), subjective JS (0.58), creating opportunities for staff development (motivation, evaluation, promotion) (0.57), information and communication within the company (0.57), supervisors managing style (0.44), the organization of work in the company (0.43), relations between employees (0.20). It seems that in the companies where IT users were mostly satisfied with their job, at the same time the OCB were observed.

Thus, null Hypothesis H01 seems to be supported.

OCB-I were associated positively and even strong enough with global JS (0.63). The strongest correlations were observed with: representing the interests of the employee (071), subjective job satisfaction (0.61), creating opportunities for staff development by motivation. evaluation and promotion (0.57), information and communication within the company (0.56), supervisors managing style (0.45), the organization of work in the company (0.43). The most weak correlation concerned relations was observed between employees (0.20).

Thus, null Hypothesis H02 seems to be supported.

This results suggest that IT users who were mostly satisfied with their jobs more often applied and perceive the ideas of OCB-I.

Also OCB-O scale was positively associated and even strong enough with JS (0.56). Again the strength of positive dependency was concerned with representing the interests of the employee (0.59) and then with information and communication within the company (0.54), creating opportunities for staff development by motivation. evaluation and promotion (0.53), subjective JS (0.53), supervisors managing style (0.41), the organization of work in the company (0.41) and in the end with relations between employees (0.20).

Thus, Null Hypothesis H03 seems to be supported.

4   Conclusions, Discussion and Future Research

The answer to first research question, that JS of IT users has a significant, positive effect on their OCB, implicates the three most important practical suggestion for management. Supervisors should take into account the subjective level of employee satisfaction, which comprises the above mentioned components, particularly, representing the interests of employees and providing opportunities for development, information and communication within the company. The important weight of information and communication as success factors in professional development in Polish companies concerned with IT and IS was also reported by Kowal et al [9].

Regarding the second research question JS of IT users has a significant, positive effect on their OCB-I.

The authors observed a greater impact of OCB-I in comparison to OCB-O concerning the phenomenon of behavior ethos manifestation. It is important for the IS development to care for the incidence of OCB-I. IT users who were mostly satisfied with their jobs more often were directed towards ideas of OCB-Is. JS may lead to greater frequency of OCB.

The answer on the third question: JS of IT users has a significant, positive effect on their OCB-O. This result is somehow comparable to the study by Kowal and Roztocki [910] and Vittell and Davis [30] which reported that in companies where professionals are more satisfied with all dimension of the job, except pay - top management encourage high ethical standards (what includes also OCB-I and OCB-O).

Overall, our study provides several important results leading to actionable conclusions.

Our most important findings indicate that JS ethical behavior premises of IT users in transition economies. The highest level of OCB is observed in organizations where the IT users really feel JS.

The authors confirmed the impact of JS on OCB the fact that IT users working more intensively in IS significantly differ from all other employee groups, what may be concerned with their higher competences and expectations. Thus, the employee satisfied with the work manifest OCB.

Current research concern with small businesses, while cited studies [9, 10] - small, medium and large companies. The authors observed that persons employed in smaller companies are more enjoyed with job, and often they are people who indeed have high competence in the use of IT, but are not professionals, such for example as IT programmers. Individuals who have higher IT competences are more satisfied with job than others.

The respondents who were less enjoyed with their job more rarely observed in their organization ethical OCB. The results were comparable to some conclusions of Vittel and Davis [31] that indicated positive correlation between professional success, ethics and JS. Kowal and Roztocki [10] also concluded some positive relation between JS and corporate social responsibility and top management action among IT professionals in Poland. However - it is surprising that the last study of Kowal and Roztocki [10] showed negative correlations between ethical optimism and satisfaction with co-workers, work itself, supervision, satisfaction with pay and promotion.

The authors claimed similarity of the OCB-I level between IT users and other employees. The results of current study are comparable to Williams and Anderson [32], who described some typical characteristic of individual behaviors of employees like: altruism, helping, courtesy, cooperative behavior and interpersonal facilitation.

The limitations of current studies concern: analyzing IT users and other employees only in Poland, variables like gender, age, position or economy sector were not carried out. The authors plan to consider these aspects in future studies.

The results of our study may be addressed to It users and professionals, especially to HR staff. Taking into account the components of JS in incentive systems increases the OCB. The applying of our conclusions may improve social and economic work effects. Applying and popularizing the ideas of OCB may be a factor of development of IS staff because ethical attitudes and behaviors increase commitment, efficiency and leads to economic growth.

References

1. Bateman, T.S., Organ, D.W.: Job Satisfaction and the good Soldier: The Relationship between Affect and Employee Citizenship. Academy of Management Journal 26, 587–595 (1983)
2. Durniat, K.: Polish adaptation of L. Rosenstiel and R. Boegel's organizational climate diagnosis questionnaire. Polish Journal of Applied Psychology 10(1), 147–168 (2012)
3. Eisele, P., D'Amato, A.: Psychological climate and its relation to work performance and well-being: The mediating role of Organizational Citizenship Behavior (OCB). Baltic Journal of Psychology 12(1.2), 4–21 (2012)
4. Ghazzawi, I.: Gender Role in Job Satisfaction: the Case of the U.S. Information Technology Professionals. Journal of Organizational Culture. Communications & Conflict 14(2), 1–34 (2010)
5. Ilies, R., Fulmer, I.S., Spitzmuller, M., Johnson, M.D.: Personality and Citizenship Behavior: The Mediating Role of Job Satisfaction. Journal of Applied Psychology 94(4), 945–959 (2009)
6. Konovsky, M.A., Organ, D.W.: Cognitive Versus Affective Determinants of Organizational Citizenship Behavior. Journal of Applied Psychology 74(1), 157–164 (1989)
7. Konovsky, M.A., Organ, D.W.: Dispositional and contextual determinants of Organizational citizenship behavior. Journal of Organizational Behavior 17, 253–266 (1996)
8. Kowal, J., Węgłowska-Rzepa, K.: The Methodological Aspects of Creating the New Research Method Based on the Choosing of Pictures and the Analysis of Creative and Recreative Functions of the Narrated Stories, Gospodarka, Rynek, Edukacja, vol. 11, pp. 65–66 (2006)
9. Kowal, J., Kwiatkowska, A., Kowal, W.: IT Project Management in Relation to Employees' Competence in Poland. In: Despres, C. (ed.) Proceedings of the 7th European Conference on Management Leadership and Governance: SKEMA Business School, Sophia-Antipolis, France, October 6-7, pp. 216–226. Academic Publishing Limited, Reading (2011)
10. Kowal, J., Roztocki, N.: Organizational Ethics and Job Satisfaction of Information Technology Professionals in Poland. In: Proceedings of the Nineteenth Americas Conference on Information Systems (AMCIS), Chicago. Illinois, USA, August 15-17 (2013)
11. Kuraś, M.: Information System – Information Technology. What Is Different than The Name of The Two Objects? [In Polish: System informacyjny – system informatyczny. Co poza nazwą różni te dwa obiekty?] (2006), http://ki.ae.krakow.pl/~kurasm /artykuly/SI-vs-SIT.pdf
12. Lee, K., Allen, N.J.: Organizational citizenship behavior and workplace deviance: The role of affect and cognitions. Journal of Applied Psychology 87(1), 131–142 (2002)
13. Chen, L.-C., Niu, H.-J., Wang, Y.-D., Yang, C., Tsaur, S.H.: Does Job Standardization Increase Organizational Citizenship Behavior? Public Personnel Management 38(3), 40–48 (2009)
14. Locke, E.A.: What is job satisfaction? Organ. Behav. Hum. Perform. 4, 309–336 (1969)
15. MacKenzie, S.B., Podsakoff, P.M., Ahearne, M.: Some possible antecedents of in-role and extra-role salesperson performance. Journal of Marketing 62, 87–98 (1998)

16. Mohammad, J., Habib, F.O., Adnan, M.: Job satisfaction and organizational citizenship behavior: an empirical study at Higher learning institutions. Asian Academy of Management Journal 16(2), 149–165 (2011)
17. Organ, D.W.: Organizational citizenship behavior: The good soldier syndrome. Lexington Books, Lexington (1988)
18. Organ, D.W.: Organizational citizenship behavior: It's construct clean-up time. Human Performance 10, 85–97 (1997)
19. Organ, D.W., Podsakoff, P.M., Mackenzie, S.B.: Organizational Citizenship Behavior. Its nature, Antecedents. Foundation for organizational science, an Consequences. A Sage Publication Series (2006)
20. Podsakoff, P.M., MacKenzie, S.B., Paine, J.B., Bachrach, D.G.: Organizational citizenship behavior: A critical review of the theoretical and empirical literature and suggestions for future research. Journal of Management 26, 513–563 (2000)
21. Rosenstiel, L., Boegel, R.: Betriebsklima geht jeden an Bayerischen Staatsministerium für Arbait. Familie und Sozialordnung, Monachium (1992)
22. Roztocki, N., Weistroffer, H.R.: Information Technology in Transition Economies. Journal of Global Information Technology Management 11(4), 2–9 (2008)
23. Schnake, M.: Organizational citizenship: A review proposed model and research agenda. Human Relations 44, 735–759 (1991)
24. Schwab, D.P., Cummings, L.L.: Theories of performance and satisfaction: a review. Industrial Relations 9, 408–430 (1970)
25. Schwepker, C.H.: Ethical Climate's Relationship to Job Satisfaction, Organizational Commitment and Turnover Intention in the Salesforce. Journal of Business Research 54(1), 39–52 (2001)
26. Smith, C.A., Organ, D.W., Near, J.P.: Organizational Citizenship Behavior: Its Nature and Antecedents. Journal of Applied Psychology 68, 653–663 (1983)
27. Steinmüller, W.: Automated Information Systems in the Private and Public Administration [In Polish: Zautomatyzowane systemy informacyjne w administracji prywatnej i publicznej.] Organizacja Metoda Technika, Nr. 1977/9 (1977)
28. Tang, T.L., Foote, D.A.: Job Satisfaction and Organizational Citizenship Behavior: Does Team Commitment Make a Difference in Self-Directed Teams?". Journal of Management Decision 46(6), 933–947 (2008)
29. Tiwana, A., Mclean, E.R.: Expertise Integration and Creativity in Information Systems Development. Journal of Management Information Systems 22(1), 45–83 (2005)
30. Vitell, S.J., Davis, D.L.: The Relationship between Ethics and Job Satisfaction: An Empirical Investigation. Journal of Business Ethics 9(6), 489–494 (1990)
31. Westfall, R.D.: An Employment-Oriented Definition of the Information Systems Field: An Educator's View. Journal of Information Systems Education 23(1), 63–70 (2012)
32. Williams, L.J., Anderson, S.E.: Job Satisfaction and Organizational Commitment as Predictors of Organizational Citizenship and In-Role Behaviors. Journal of Management 17, 601–617 (1991)
33. Xia, W., Lee, G.: Complexity of Information Systems Development Projects: Conceptualization and Measurement Development. Journal of Management Information Systems 22(1), 13–43 (2005)

Do Distracting Dashboards Matter?
Evidence from an Eye Tracking Study

Palash Bera

John Cook School of Business, Saint Louis University, 3674 Lindell Boulevard,
Saint Louis, Missouri, 63108, USA
pbera@slu.edu

Abstract. This paper analyzes the effect of distraction on using dashboards. Distraction in dashboards was introduced by increasing the non-pixel data and not highlighting the task relevant area. Eye tracking technology was used to precisely measure how much time and effort users make to understand the dashboards. Eye tracking technology was used as it provides the details of the mental processes by which users make decision regarding the tasks. A laboratory study was conducted using eye tracking technology to understand how subjects use two types of dashboards- distracted and non-distracted to perform certain tasks. Results show that although both distracted and non-distracted groups performed equally well in answering the tasks, the distracted group had significantly high overall fixation time and count. This shows that the distracted group spent more time and effort in answering the task. Also, it was found that the non-distracted group spent more time and effort on the specific area of the dashboard where the answer of the task was available.

Keywords: Business Intelligence, Dashboard Design, Eye Tracking.

1   Introduction

Business Intelligence (BI) covers strategies and technologies to achieve knowledge about status, potentials, and perspectives of an organization from multiple data sources. A key BI tool that is getting popular in today's business world is – *dashboard*. A dashboard is a visual display of the most important information necessary to achieve one or more objectives, consolidated and arranged on a single screen [1]. Dashboards help users visually identify trends, patterns, and anomalies and reason about what they see and help guide them towards effective decision making [2]. As data visualization is becoming more important, a key aspect of dashboard is its design. In this paper, one element that might affect dashboard design is considered-*distraction*. The focus of this paper is to understand how dashboards designed with distracted elements affect users' performance and effort in using the dashboards for certain tasks.

S. Wrycza (Ed.): SIGSAND/PLAIS EuroSymposium 2014, LNBIP 193, pp. 65–74, 2014.

2   Background

2.1   Distraction in Dashboards

The effect of distraction on search task has been extensively studied in psychology literature. In particular, the effect of distractors such as color on visual search tasks has been investigated [3]. In this paper, distraction is introduced in dashboards by using the data-ink ratio concept. The basis of data-ink ratio concept is when quantitative information is used to display information in printed form, then some of the ink is used to present data (such as table) and some of the ink is used to present non-data (such as graph). Tuffe [4] suggests that a large share of the ink should be used to represent data-information and less on non-data information. Few [5] modified this principle by suggesting that as dashboards are always displayed on computer screen, so the word ink should be replaced by pixel. Thus he suggests that "across the entire dashboard, non-data pixels- any pixels that are not used to display data, excluding a blank background-should be reduced to a reasonable minimum (page, 97, [5])."

In this research, distraction in dashboard is introduced by: (1) using a dashboard title that is irrelevant to the task (2) using a picture that is unnecessary for the task and (3) manipulating the position of information that is relevant to the task. The first two elements of distraction emphasize the use of non-data pixels as both pictures and texts are unrelated to the task. By doing so the area of non-data pixel increases in the dashboard. The third element places low emphasis to the information required to perform the task.

Fig. 1. Example of a distracted Dashboard

To provide an example of a distracted dashboard, consider the dashboard in Figure 1 below. In this dashboard, the specific task that user needs to perform is to answer - in how many months, the number of closed sales was exactly the same as the number of pending sales? To answer this task, a user needs to refer to the lower part of the dashboard and calculate the number of times the blue peaks of the histogram are touched by the orange line. The title "analysis of real estate sales" and the picture of the house for sale are not necessary here as they distract the user and do not relate to the task. Moreover, the answer of the task lies in the lower part of the dashboard (histogram and line chart) and not in the upper part of the dashboard. Therefore the graph of months of supply of homes is causing distraction to the user as the relevant information is not present here.

In comparison to Figure 1, Figure 2 shows a non-distracted dashboard where the three elements of distractions are not used. Moreover, the information required to perform the task is presented at the top part of the dashboard.

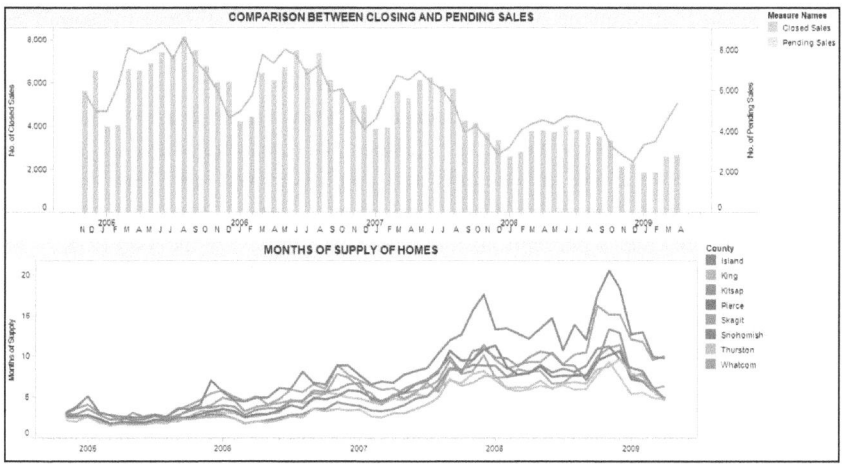

Fig. 2. Example of a distracted Dashboard

To understand how the users perform tasks using the dashboards, an eye tracking study was conducted with students as subjects. Eye tracking technology was used to precisely measure how much time and effort users make to perform a task using the dashboards. Also, analysis of how much attention one spends on particular area of the dashboard can give us the details of the mental processes by which users came to the decision regarding the tasks. Without use of such technology, it is difficult to find the effect of the dashboards as the difference in the two dashboards design is subtle. The next section discusses eye tracking technology.

2.2   Eye Tracking Technology

Eye-tracking enables researchers to measure subjects' eye movements while reading a text or viewing a picture. Involuntary and voluntary eye movement responses reflect the internal processing of information[6]. Eye tracking offers a window into how individuals read and scan information that is displayed to them. During decision making tasks where users are allowed to view the decision information, eye movements can be considered to provide a valid measure of distribution of attention [7]. By relating eye movement behavior with decision making data, one can obtain a comprehensive picture of the decision making process. Eye-tracking uncovers how decision making takes place and provides insight into the content of information processing.

Jacob [8] explains the physiology behind the eye movements. An area termed fovea is located near the center of the retina and is densely covered with receptors. Fovea provides high acuity (near) vision than the peripheral vision. Peripheral vision is generally inadequate to see an object clearly and generally requires that the object to be viewed foveally. When an object is viewed foveally by a reader, he/she makes attempt to understand the content. Therefore, for eye tracking purposes, in order to see an object clearly, one must move the eyeball to make that object appear directly on the fovea. At that moment, a person's eye position can be directly measured by an eye tracker.

Research in tracking eye movements is common in psychology and physiology. Much of the early work on these areas focused on how human eyes operate and what it can reveal about perceptual processes [9]. Recently researchers have begun to focus on the relation between eye movements and cognitive processes. Eye tracking has been used in improving graphics design and website design (e.g. [10]and [11]) and understanding consumer's perceptions and attention of advertisements ([12]).

Although eye-tracking technology has been used for over 30 years, the technology was unreliable and data interpretation was time consuming [13]. Over the years, the technology has become more reliable, user friendly, and affordable [9] and thus suitable for analyzing mental processes of users. This technology allows us to objectively and precisely measure users' attention to specific information contained in the dashboard. This further allows us to understand the actual process by which this information is acquired and how it is used to understand a dashboard.

Two common eye movement metrics are: eye fixations and eye saccades [14]. Eye movements are made up of short bursts of stationary visual display termed fixations and are filled up with rapid and continuous movements termed saccades [8]. During fixations, eyes remain almost motionless, whereas saccades are movements from one fixation to another. A typical fixation lasts for 200-300 milliseconds approximately and is generally understood to indicate where viewer's attention is directed [6]. When eyes fixate on a certain area, the brain starts to process the visual information received from the eyes [6].

3   Empirical Study

To test the effect of distraction on BI dashboards, a laboratory study was conducted using eye tracking technology. In this section, the hypotheses, subjects, tasks, design, and procedure of this study are discussed.

3.1   Hypotheses

Two relevant theories that help to build the hypotheses are: (a) cognitive task fit theory and (b) cognitive overload theory. Cognitive task fit theory [15] states that different representations of information are suitable for different tasks and different audiences. Task performance improves if there is a cognitive fit among form of representation, task characteristics, and problem solvers skills. Cognitive overload occurs when a viewer has more information to process than his/her available working memory [16]. With the distracted dashboard, there is a misfit between the task and representation (dashboard) and this might affect the performance of the task. Also because of the picture and the title, there can be cognitive overload to the user. Fixation movements can reveal the amount of information that is processed. Research has demonstrated that information complexity, task complexity, and familiarity of visual display will influence fixation duration [17]. As distraction is expected to increase information complexity in the entire dashboard thus it can be expected that fixation time and fixation counts will be higher for subjects viewing the distracted dashboards compared to the non-distracted dashboards. Accordingly,

H1: Subjects provided with distracted dashboards will have higher overall fixation number and fixation time compared to the subjects provided with non-distracted dashboard.

Apart from analyzing the eye movements of the entire dashboards, it is also possible to analyze the eye movements of specific parts of the dashboard. Researchers define "areas of interest" (AOI) over certain parts of a display and analyze the eye movements that fall within such areas. In this way, the visibility, meaningfulness, and placement of specific elements can be objectively evaluated [18]. For tasks that can be answered from specific parts of the dashboard, it is expected that these parts will be used more frequently than other parts of the dashboard. However, as users of non-distracted dashboards can identify these parts or relevant areas more easily, therefore it is expected that such users spend more time and effort in these relevant areas of the dashboard. Accordingly,

H2: Subjects provided with non-distracted dashboards will have higher fixation number and fixation time on viewing task relevant areas compared to the subjects provided with distracted dashboards.

3.2   Subjects and Tasks

30 MIS graduate students of a Southern US University were recruited as subjects in this study. These students were enrolled in IS Analysis and IS Design courses. The students also took graduate statistics courses and thus were familiar with the elements of dashboards such as graphs and tables. A small number of subjects is typical in eye tracking studies as each subject has to participate in the study one at a time. For participation, subjects were awarded course credits. Subjects were asked to answer questions based on the dashboards. To generalize the results, two sets of dashboards were used. The first set of dashboards (Figures 1 and 2) is based on a real estate

domain and the second set is based on coffee sales domain (Figure 3). The two tasks for each dashboard can be answered by viewing a specific area of dashboard. For the distracted dashboard, the relevant area was in the bottom part where as for the non-distracted dashboard the relevant area was on the top part of the dashboard. The two tasks that were used are (1) in how many months, the number of closed sales was exactly the same as the number of pending sales? (real estate domain) and (2) in which two months, budget profit was higher than budget sales? (coffee sales domain).

3.3   Design and Procedures

The study had a between-group design were subjects were randomly assigned to one of the variations (distracted and non-distracted) of the dashboard. 15 subjects were provided with the distracted dashboards and the other 15 were provided with the non-distracted dashboards. The subjects answered two tasks related to two dashboards, while their eye movements were tracked. Prior to this, their eyes were calibrated and validated (a standard procedure for eye tracking) by asking them to follow a series of dots in the screen. After calibration, they were shown a task in a screen and asked to read the task carefully. Then they pressed a joystick to see the dashboard and verbalized the answer. This was done to avoid eye movements associated with writing the answers. This sequence was repeated for another dashboard. Eye movements were recorded by the EyeLink 1000 eye tracking software. The verbalizations were also recorded. The tracker records a minimum fixation of 4 milliseconds.

4   Analysis

The percent of correct answers of the tasks for the distracted group was 87 and for the undistracted group was 89 (average value for two domains). Thus in terms of answering tasks from the dashboards, there was no statistical significant difference between the two groups. However, the eye movement data revealed a different pattern. The fixation duration (in seconds) and the number of fixations were compared between the two groups for both tasks. Fixation count reveals the amount of cognitive processing [19]. Specifically, a large number of fixations on a large area (such as overall dashboard) indicates less efficient search possibly resulting from a poor arrangement of display elements [9]. Table 1 shows that the subjects who viewed the distracted dashboards spent significant more time and had more number of fixation count than those who viewed the non-distracted dashboards (at a 5% level of significance). This result supports H1 and suggests that subjects had to put more cognitive efforts when using distracted dashboards compared to the subjects using non-distracted dashboards.

Table 1. Analysis of eye movements of overall dashboards for both groups

Overall Dashboard	Distracted (N = 15)		Non Distracted (N = 15)			
Domain: Coffee Sales						
	M	**SD**	**M**	**SD**	**t**	**p**
Duration (in sec.)	28.435	10.35	22.44	4.23	2.07	0.03*
No. of Fixations	104.13	32.57	78.66	17.54	2.66	0.01*
Domain: Real Estate						
Duration (in sec.)	61.04	24.12	44.57	17.16	2.15	0.02*
No. of Fixations	199.67	79.94	139.00	57.10	2.39	0.01*

M- Mean, SD- Standard Deviation, * Significant at 5% level

Next, the eye movements were compared between groups for the task relevant area of the dashboard. As only one area was relevant for the task in the dashboard so the eye movements were compared between groups for this specific area only. Instead of using the total number of fixation and time for the specific area, percentage time duration and percentage fixation count were used for high accuracy. This percentage was calculated using the total fixation (number or time) of the specific area divided by the total fixation (number or time) of the entire dashboard. Table 2 presents the results. For example, 0.61 in line 4 of Table 2 means that the mean percentage of fixation duration (%OfFixationDuration) for the distracted group looking at the relevant area of coffee sales dashboard is 61%. Table 2 supports H2 and shows that the non-distracted group spent significant more time and effort in using the task relevant area compared to the distracted group (at a 5% level of significance).

Table 2. Analysis of eye movements of task relevant area of dashboards for both groups

Task relevant area	Distracted (N = 15)		Non Distracted (N = 15)			
Domain: Coffee Sales						
	M	**SD**	**M**	**SD**	**t**	**p**
%OfFixationDuration	0.61	0.10	0.75	0.07	4.30	0.00*
%OfFixations	0.62	0.11	0.73	0.09	2.86	0.00*
Domain: Real Estate						
%OfFixationDuration	0.80	0.04	0.89	0.02	3.28	0.00*
%OfFixations	0.78	0.05	0.87	0.09	3.15	0.00*

M- Mean, SD- Standard Deviation, * Significant at 5% level

Sample heat maps from individual subjects (Figures 3a and 3b) showing the intensity of the fixation count (darker means higher fixation count) confirms that subjects in the distracted group spent more time in looking areas that are not related to the task (Figure 3a). This shows that distraction does have an effect in terms of time and effort even if the tasks performed by the subjects are not complex.

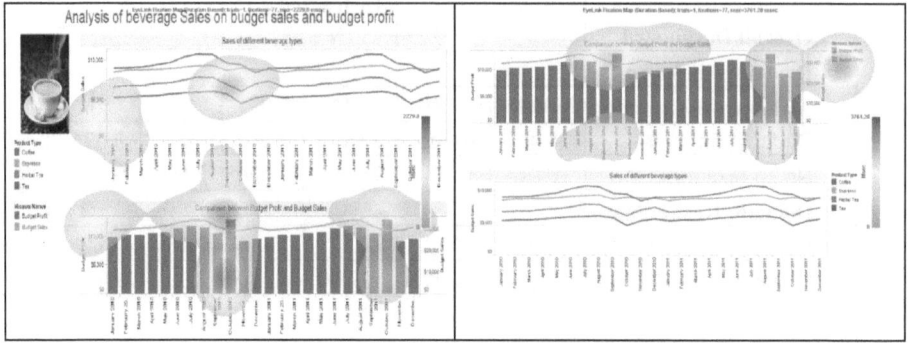

Fig. 3a. Sample heat map of a distracted Dashboard

Fig. 3b. Sample heat map of an non-distracted Dashboard

5   Future Work and Conclusion

This research analyses the effect of distraction on using dashboards. Distraction was introduced by increasing the non-pixel data and not highlighting the task relevant area. Results show that although both groups performed equally well for specific tasks, the distracted group had significantly high overall fixation time and count. This shows that the distracted group spent more time and effort in answering the task. Also, it was found that the non-distracted group spent more time and effort on the specific area where the answer of the task was available.

This study was conducted with two limitations. First, the task that was asked to the subjects was direct and could be easily answered if subjects focused on one part of the dashboard. Thus the true effect of distraction was not identified. Future tasks are planned that will be more complicated and will require more attention. Such task can be subjective in nature and might require integration of multiple areas of the dashboard (e.g. in real estate domain –which county could be ideal in terms of buying a house in 2010?). Perhaps distraction will affect the performance of such complex task. It seems that although there is an effect of distraction for the task that was asked but the task was simple and could be easily answered even with distraction. Second, the eye tracker Eyelink 1000 has a specific limitation in terms of interactivity. The tracker does not allow subjects to go back and forth with the task and the dashboard (i.e. eye movements are recorded in only one screen). Thus subjects were forced to read the task carefully before looking at the dashboard. A more complex task will require subjects to go back and forth with the task and the dashboard. Moreover, the eye tracker treated the dashboards as static pictures. But as dashboards are interactive (e.g. clicking on one element of the dashboard affects other) therefore the setting of the experiment was not realistic. To answer a complex question, users need to interact with the dashboard. For example in the real estate dashboard, users should be able to select one county and the rest of the dashboard will display data related to the selected county. To overcome the software limitations, the planned experiment will use Tobii X 60 – an advanced eye tracker that records eye movements while allowing dashboards to be interactive and also allows subjects moving back and forth between the dashboard and the task.

It is also important to note that there was no difference in task performance between the two groups. This shows that the difference was subtle and to identify the difference in the two dashboards, eye tracking is an appropriate technique. We also need to be aware that in this study, although it was possible to identify that the distracted dashboard creates more overload than the non-distracted one, the mental processes on how the users perform tasks were not identified using the eye tracking technique.

References

1. Few, S.: Dashboard Confusion. In: Intelligent Enterprise (2004)
2. Brath, R., Peters, M.: Dashboard Design: Why Design is Important. DM Direct Newsletter (2004)
3. Bekkering, H., Neggers, S.: Visual Search is Modulated by Action Intentions. Psychological Science 13(4), 370–374 (2002)
4. Tuffe, E.: Envisioning Information. G. Press (1990)
5. Few, S.: Information Dashboard Design: Displaying data for at-a-glance monitoring. Analytics Press (2012)
6. Rayner, K.: Eye movements in reading and information processing: 20 years of research. Psychological Bulletin 124(3), 372–422 (1998)
7. Glaholt, M.G., Reingold, E.M.: Eye movement monitoring as a process tracing methodology in decision making research. Journal of Neuroscience, Psychology, and Economics 4, 125–146 (2011)
8. Jacob, R.: Eye Tracking in Advanced Interface Design. In: Virtual Environments and Advanced Interface Design. Oxford University Press, New York (1995)
9. Jacob, R., Karn, K.: Eye tracking in human-computer interaction and usability research: Ready to deliver the promises. In: Hyona, R. (ed.) The Mind's Eye: Cognitive and Applied Aspects of Eye Movement Research. Elsevier, Oxford (2003)
10. Chu, S., Paul, N., Ruel, L.: Using eye tracking technology to examine the effectiveness of design elements on news websites. Information Design Journal 17, 31–43 (2009)
11. Canham, M., Hegarty, M.: Effects of knowledge and display design on comprehension of complex graphics. Learning and Instruction 20, 155–166 (2010)
12. Lam, S.Y., Chau, A.W.-L., Wong, T.J.: Thumbnails as online product displays: How consumers process them. Journal of Interactive Marketing 21, 36–59 (2007)
13. Collewijn, H.: Eye movement recording. In: Carpenter, R.H.S., Robson, J.G. (eds.) Vision Research: A Practical Guide to Laboratory Methods, pp. 245–285. Oxford University Press (1999)
14. Sharif, B., Maletic, J.: An eye tracking study on the effects of layout in understanding the role of design patterns. In: IEEE International Conference on Software Maintenance (2010)
15. Vessey, I., Galletta, D.: Cognitive Fit: An Empirical Study of Information Acquisition. Information Systems Research 2(1), 63–84 (1991)

16. Mayer, R.E.: Human Nonadversary Problem Solving. In: Gilhooly, K.J. (ed.) Human and Machine Problem Solving. Plenum Press, New York (1989)
17. Duchowski, A.T.: A Breadth-First Survey of Eye Tracking Applications. Behavior Research Methods, Instruments, & Computers 34(4), 455–470 (2002)
18. Goldberg, H.J., Kotval, X.P.: Computer interface evaluation using eye movements: Methods and constructs. International Journal of Industrial Ergonomics 24, 631–645 (1999)
19. Poole, A., Ball, L.J.: Eye Tracking in Human-Computer Interaction and Usability Research: Current Status and Future Prospects. In: Ghaoui, C. (ed.) Encyclopedia of Human Computer Interaction, pp. 211–219. Idea Group, Hershey (2006)

Models of Research Activity Measurement: Web-Based Monitoring Implementation

Olga Cherednichenko[1], Olha Yanholenko[1], Olena Iakovleva[2], and Oleksii Kustov[1]

[1] National Technical University "Kharkiv Politechnic Institute", Frunze str. 21,
61002 Kharkiv, Ukraine
[2] Kharkiv National University of Radioelectronics, Lenina str. 14,
61166 Kharkiv, Ukraine
{olha_cherednichenko,olga_ya26,helen_yakovleva}@mail.ru,
a.kustov@ct-college.com.ua

Abstract. The given work is devoted to the development of models that form the basis of measurement activities in web-based monitoring and evaluation information system of university, which is a part of quality management system. The suggested models include the model of higher education establishment representing the properties of key business processes, the data sources model giving the description of allocation of relevant data on the web, the scoring model supporting those data transformation into scores, and the measurement model giving the mechanism of scores processing into the values of specified properties. In the given work we suggest to use Rasch model as a measurement model. The case study concerns the university research process. Our experiment represents the measurement of the level of research results publishing activity. The obtained results meet the requirement of reliability and can be used in management purposes.

Keywords: web-based monitoring, higher education, research process, measurement model, Rasch model, information system.

1 Introduction

Higher education is a complex social sphere where the main role is played by universities. Higher education establishment (HEE) in the modern conditions has to struggle for new prospective students. That's why universities pay much attention to their reputation and favorable position among competitors. The results of HEE functioning determine its success on the market of educational services. To control the results HEE needs the support of quality management system.

The basic directions of university work include educational and research activities. The quality of education can be considered from different points of view. Many authors treat education quality as the level of knowledge and skills obtained by students during the studying. Such approach is analyzed in psychometric theory [1]. Considering education as a process leads to its analysis based on Total Quality Management approach [2]. Education quality as a quality of service can be considered

S. Wrycza (Ed.): SIGSAND/PLAIS EuroSymposium 2014, LNBIP 193, pp. 75–87, 2014.

from stakeholders points of view. For example, our previous research was devoted to evaluation of students' satisfaction with the obtained education quality [3]. The available resources of HEE also determine the quality of provided services. Therefore we solved the problem of resources evaluation [4]. Since the quality category is very complex and combines the values of different object's features, we often deal with the problem of comprehensive assessment [5].

Research activities of HEE also require special analysis. The quality of research results influences the position of university in scientific society. We suggest the approach of evaluation of the level of activity in conference organization as one of possible indicators of the research results [6]. In the given work we provide an example of measurement of the level of HEE activities in research results publishing, which is another possible indicator of research results quality.

Such kind of evaluation and analysis should be conducted on a constant basis. This is the problem of HEE quality management system. To bring useful results to the HEE management this system must provide it with the actual data on educational and research results permanently. This problem can be solved within the monitoring and evaluation process. Education quality monitoring is continuous collection of data on education quality indicators used to provide management of HEE and main stakeholders with indications of objectives achievement [7]. As a rule, monitoring is complemented by evaluation, which implies the assessment of the present outcomes against the planned results. In our research we consider results-based monitoring and evaluation as the one which is oriented on goals and results of the system functioning [8]. Moreover, we develop the framework of web-based monitoring that uses the data stored on the web [9].

Analyzing other possible ways to obtain evidences about HEE results of functioning we can address to university rankings [10]. However, usage of different rankings with management objectives is limited due to several reasons. First of all, the methodology of ranking is not always available; therefore we may not know the way how the positions of universities are determined. This means that ranking results cannot be properly used, since the management does not know exactly which aspects of HEE work must be improved. Secondly, the analysis of the most recognized rankings allows to conclude that the quality of research (which is the subject of this work) is considered from the point of view of citation indices, academic reputation, research income and output. But this is not quite enough. There are other aspects of research activities that should be taken into account.

Compared to university ranking the suggested approach of web-based monitoring overcomes the described issues. Additionally we can specify its following advantages: the possibility to collect data automatically and on the permanent basis; the frequency of estimates obtaining is defined by HEE management, but not by ranking organization; the scale of HEE estimates are not in the scale of order (like in rankings), but in the interval scale.

We suppose that the results of HEE work are reflected not only in its own official documents or in the documentation of external organizations. Our basic hypothesis is that we can find the evidence of university's functioning on different web sites, for example, conference and journal web sites, news and blogs, social networks, etc. [7]. This data must be collected from different sources and analyzed.

We narrow the scope of our interest by the domain of university research process. This process is investigated in the given paper in a detailed manner. We suggest the approach of measurement of research activities results based on the data stored on the web. When talking about continuous process of collection and processing of big volumes of data on the web, it is obvious that such problem should be solved with the help of the automated tool. Therefore the given work is devoted to the development of models supporting research activity monitoring and measurement, which form the basis of corresponding information system.

The rest of this paper is organized in the following way. Section 2 describes research business process in HEE. Section 3 represents the design assessment models of research activities. The case study is given in Section 4. The conclusions and future work are presented in Section 5.

2   Research Business Process in HEE

In order to determine the data sources for monitoring of research activities in university we suggest to discuss this process in details (fig. 1). A research process is initiated by some interested scientist or a group of scientists and is supported by the HEE where they work. A research process takes the form of some project which may try to get the grant or may start without additional funding. Research activities include analysis of the problem, literature review, experiments, surveys, interviews, case studies.

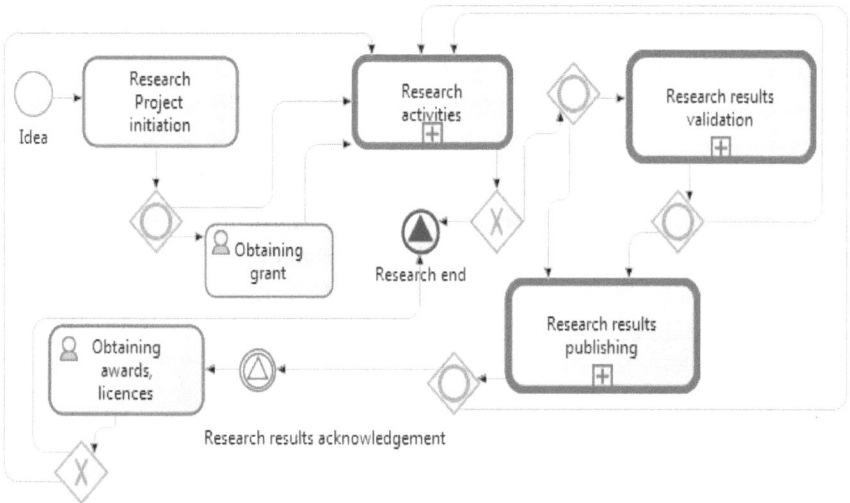

Fig. 1. Research business process

The obtained results should be validated. The first step of validation is presentation of ideas and their discussion on the conferences. It is important to get both denial and acceptance of papers after the reviewing process. In the given work we use the notion "validation" in the sense of research results presentation on the conferences, and not in any other meaning. After this the approved research results can be published in the journals and introduced in industrial practice. The achievements of the authors can be recognized and acknowledged by scientific society which may lead to obtaining of awards and licenses. In any case a research continues until its goal is achieved. So the results of research can be available for public via participation in conferences and publishing them in journals.

Research results validation is presented on fig. 2 in DFD notation. The contribution of authors in this process is a written paper, while the university expresses its will to organize the conference and provides necessary resources. The outcomes of research results validation are the camera-ready versions of papers collected in the conference proceedings and the Call for Paper allocate on the conference web site and distributed over scientific society.

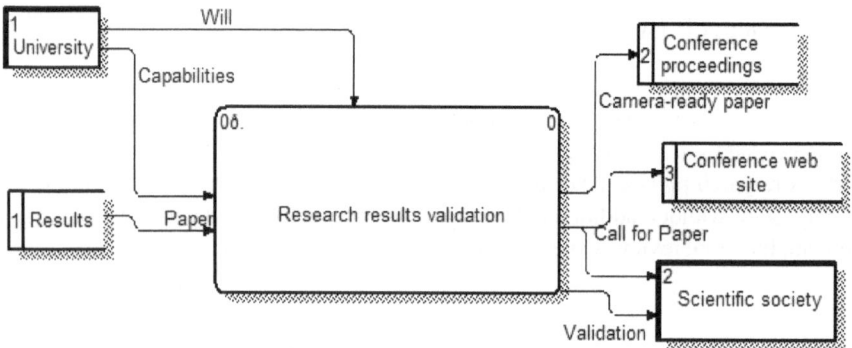

Fig. 2. Research results validation process

Research results validation is decomposed into conference organization, participation in it and carrying out (fig. 3). Conference carrying itself does not bring much to monitoring of research results validation, except for the possible awards of papers, which were successfully presented on the conference.

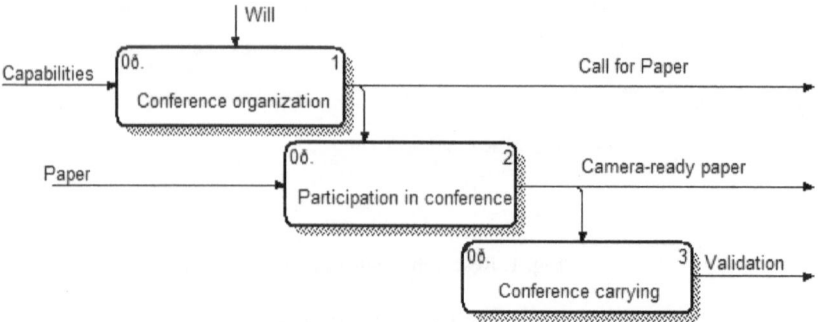

Fig. 3. Research results validation – decomposition

Conference organization starts from Organizing committee forming (fig. 4). Having capabilities provided by university, the members of this committee develop a program of event. Topic areas covered by conference program determine the choice of invited members of Program committee. All necessary information is published in the form of Call for Paper, which is allocated on the conference web site. The promotion of the conference implies the distribution of information about it in the scientific society.

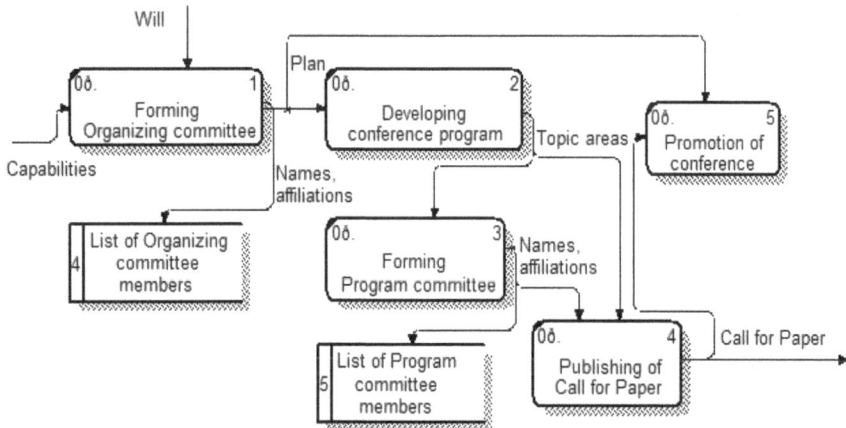

Fig. 4. Conference organization – decomposition

Participation in a conference means that an author submits a prepared paper. After this the reviewing takes place (fig. 5). In the case of acceptance an author is asked to prepare a final camera-ready version of the paper, which is published in conference proceedings.

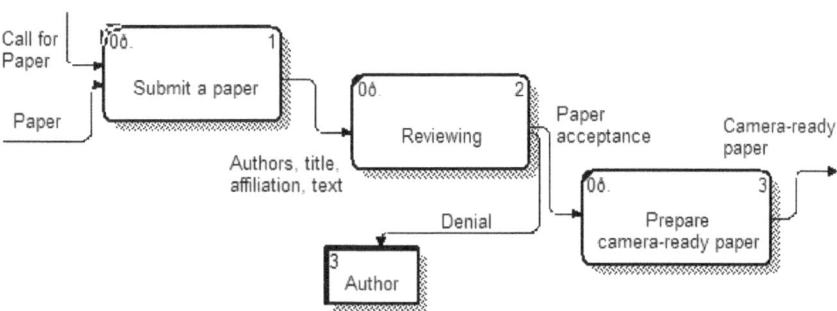

Fig. 5. Participation in conference – decomposition

Analyzing the data flows, we can conclude that information about results of HEE research activities can be found on the conference web site. In particular, the data about Organizing and Program committees, topic areas, important dates can be found in Call for Paper. Authors' names, their affiliations, papers' abstracts or even full texts are usually allocated on the conference web site as well as the Call for Paper document. This data is useful for the web-based monitoring. The sources of data for monitoring include conferences' web sites containing Calls for Paper and information about participants.

In the same manner we analyzed research results publishing process (fig. 6, 7). Its context DFD diagram is shown on fig. 6. The data about paper's authors and their affiliation can be found on the journal web site.

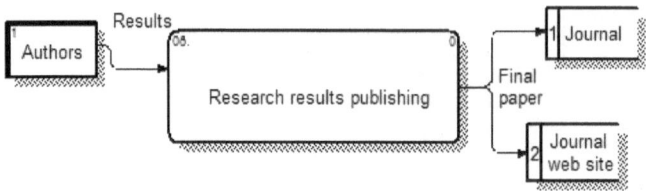

Fig. 6. Research results publishing process

The members of editorial board are involved on the stage of paper reviewing of research results publishing process (fig. 7). As a rule, the information about editorial board and affiliations are presented on the journal web site.

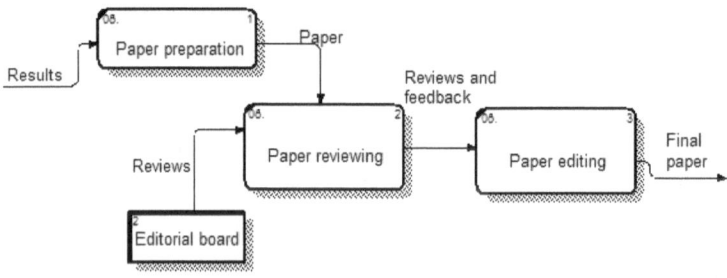

Fig. 7. Research results publishing – decomposition

Analyzing key processes or research results publishing, we concluded that the relevant data for monitoring can be found on journals' web sites. Here we pay attention to the journal founder, composition of editorial board and authors' information. The presented processes are simplified to the extent which allows us to find out the relevant information for monitoring, which can further be used in decision making.

The goal of the given work is to develop the software component that allows measuring the features of research process (such as the level of activities) based on the data collected from the web.

3   Design Assessment Models

In order to evaluate the results of a HEE we need to develop an assessment framework which represents the way of inferring based on the observed data (fig. 8). A HEE model concerns the properties of different processes that take place in HEE and which we want to explore. We denote them through $\alpha_i, i = \overline{1, N}$. These are the latent variables that reflect the results of HEE management in the specific area, such as educational process, research, etc. Our goal is to measure these unobservable variables, since we suppose that they are useful for organizing of our thinking about the HEE. Such data can be used by HEE management with the purpose of planning its development strategy.

The data sources model represents the observable evidences of HEE business processes. In the given work we consider the web as a data source where we can find the clues about university activities. These are real world situations where the results of HEEs functioning are located. In order to describe what web pages contain relevant information we need ontology. It represents the knowledge about the potential data sources for monitoring in the formal way. Such ontology must be constructed with the assistance of experts in HEEs functioning.

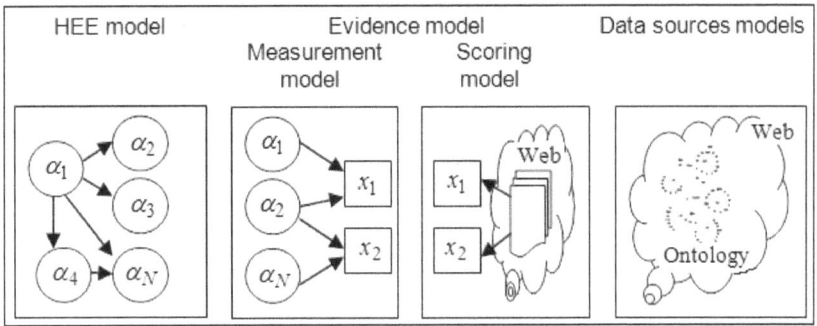

Fig. 8. Assessment framework

The evidence model describes the mechanism of inference about HEE activities levels. We denote them through latent variables α_i . The estimates of these latent variables are based on the observable data from the web. To define which data from the web pages should be extracted and how it should be interpreted we need a scoring model. This model helps to assign some categorical values (scores) that are associated with the evidences extracted from the web. The activities in which HEE takes part and which are located on the web are considered as items in this work. For each item we get the score of HEE. These scores may be expressed in different scales, for instance, in dichotomous (1/0) scale, when we are interested in the presence or absence of HEE results clues on some web page. Scoring model helps to understand in what way and with what value the evidences from the web affect our vision of the results of HEE processes.

When we have the scores, we need another model to transform them into values α_i of HEE properties. What we need is a measurement model. The calculation procedure provided by this model must give the quantitative values in a proper scale that meet the requirements of reliability and validity.

Let's consider the described models for university research business process. We would like to investigate different properties α_i of research process. They are the level of activity in conference organization, publishing and research projects participation. The set of possible indicators is given in Table 1.

Table 1. Research results indicators

Participation in conferences	Publishing of research results	Research projects
Representation in Organizing committee	Issue of journals	Participation in research projects
Representation in Program committee	Issue of collections of articles	Awards, distinctions, rewards
Venue	Representation in journal among authors	Grants obtaining
Representation among authors	Representation in journal editorial board	Licenses, patents obtaining

The data sources model is represented by the web sites where we can find the clues of university's research activities. It is described by the ontology which sets the rules of web page marking as a source of data for monitoring. For example, such rules for indicators of research results publishing include the following: a web page is a source of data, if it contains the name of journal, year, words "editorial board" in the header and the list of members with their affiliations. The estimation of a web page subject to its ability to be a data source for monitoring is conducted based on comparator identification [11] (fig. 9).

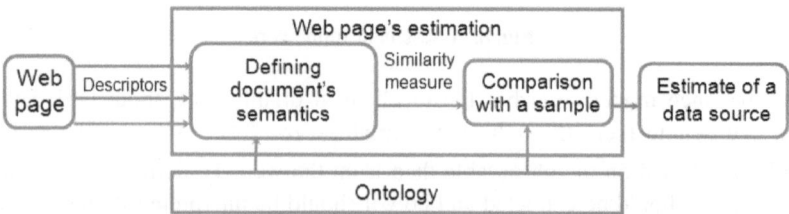

Fig. 9. Web page estimation based on comparator identification

The scoring model gives the set of items on which the HEE is scored. This can be the representation of HEE in editorial board of a journal or in conference program committee which can be scored as 1 or 0.

We can consider different measurement models that provide the values of latent variables based on the observed data. This can be the basic model of Classical Test Theory [12]:

$$X_j = \theta + E_j, \tag{1}$$

where X is the observed score of a person; θ is person's level of ability; E is an error of measurement.

The main disadvantage of this model is that it confounds latent variable measures and items characteristics. Therefore we suggest to use the models of Item Response Theory [13], in particular Rasch model with our designations. The probability of obtaining 1 score is expressed as:

$$P(x_{ij} = 1 | \alpha_i, \sigma_j) = \frac{\exp(\alpha_i - \sigma_j)}{1 + \exp(\alpha_i - \sigma_j)}, \tag{2}$$

where x_{ij} is a score i-th HEE for j-th item; α_i is a level of property affected by the research activities in i-th HEE; σ_j is a difficulty of j-th item.

Our main assumption is that the more active a HEE is in research – the higher is the probability that its results can be found on the web. This is confirmed in practice. The universities with long history and well-known scientists take higher positions in rankings and are more actively highlighted on the web.

The difference $(\alpha_i - \sigma_j)$ is considered as a single variable. When the probability P_{ij} tends to 1, the value $(\alpha_i - \sigma_j)$ is much bigger, than 0. In the opposite case, when P_{ij} tends to 0, the value $(\alpha_i - \sigma_j)$ is much less, than 0. If $P_{ij} = 0.5$, $\alpha_i = \sigma_j$. The measures of model's variables are made in logits [12].

The main advantages of Rasch model, which make it so attractive, are its interval scale and independence of variables. The interval scale provides us with comparable measurements. So we can compare the results of different HEEs or the results of the same HEE in different time periods. Variables independence allows us to obtain the value of the level of research process activities independently from the raw data extracted from the web. So we can say that if we analyze the data from the web sites of different conferences or journals, this will not affect the measurement result of the latent variable, since it remains the same and we just change the items.

The described models form the basis of the developed multiagent information system of HEE research results monitoring [14]. Its measurement agent incorporates Rasch model for indicators assessment.

4   Case Study

The case study of the given work is devoted to measurement of the level of universities activities in research results publishing. We distinguish two types of activities in publishing: participation in editorial board of HEE's employees and participation as authors of scientific papers.

As a rule, every scientific journal has its website. The information about editorial board is allocated on it. The web page with editorial board includes the names of its members and their affiliations. Also the web site contains information about published papers, their authors and affiliations. We can find the whole text of articles, abstracts or just the contents of the volume. In any case we can analyze the data about authors and their places of work.

The described HEE, data sources, scoring and measurement models are used for measuring of the level of activity of research results publishing. We examined 50 top universities in Computer Science according to QS University ranking [10]. We chose 100 top journals in Computer Science from Microsoft Academic Search ranking [15].

We consider the following set of items $j = \overline{1,6}$: j_1, j_2, j_3 reflect the presence of at least one university employee in editorial board of 100, 50 and 25 top journals respectively; j_4, j_5, j_6 reflect the presence of at least one university employee among the authors of the last volume of 100, 50 and 25 top journals respectively. If at least one HEE employee is present in the specified items, it is assigned 1, otherwise – 0. The frequency of HEEs scores is shown on fig. 10.

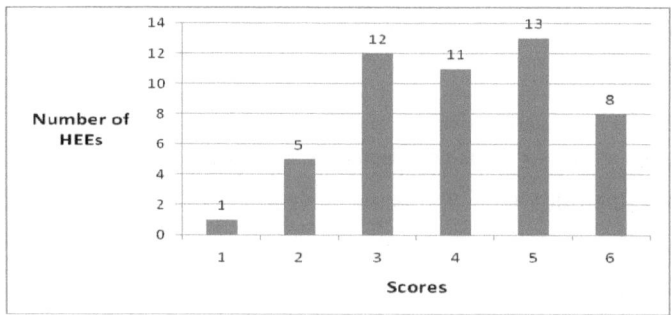

Fig. 10. HEEs scores frequency

The collected data is processed based on Rasch model (2). The results of experiment for the first 10 universities in the list are given in Tables 2-4. The total score, presence rate, initial and final estimates of α and standard errors (SE) for the first 10 universities are given in Table 2.

Table 2. Estimation of the level of activity of research results publishing (fragment)

HEE	Score	Presence rate	Initial α, logits	Final α, logits	SE
Massachusetts Institute of Technology (MIT)	5	0,83	2.33	2.54	0,46
Stanford University	5	0,83	2.33	2.54	0,46
Carnegie Mellon University	4	0,67	1.47	1.69	0,43
University of Cambridge	5	0,83	2.33	2.54	0,46
Harvard University	5	0,83	2.33	2.54	0,46
University of California, Berkeley	4	0,67	1.47	1.69	0,43
University of Oxford	4	0,67	1.47	1.69	0,43
ETH Zurich	4	0,67	1.47	1.69	0,43
National University of Singapore	4	0,67	1.47	1.69	0,43
Princeton University	5	0,83	2.33	2.54	0,46

Maximal, minimal and mean values of the level of activity in research results publishing, variance (Var), standard deviation (SD) and mean standard error are presented in Table 3.

Table 3. Analysis of the results of all HEEs

Max α, logits	Min α, logits	Mean α, logits	Var α	SD α	Mean SE
2.54	-1,12	0.65	1,13	1,14	0,47

The reliability calculations are given in Table 4, which includes separation index and separation reliability for HEEs (PSI and PSR) and items (ISI and ISR).

Table 4. Reliability estimation

PSI	PSR	ISI	ISR
1,64	0,89	1,02	0,78

Variable α_i of the level of research activities in publishing varies from -5 to 5 logits according to Rasch model. Analyzing the obtained results, we can conclude that in the most extent the values of HEEs correspond to their initial ranking. Some differences can be explained by the fact that the ranking considers many indicators, but not the single one. However, our result concerning the research publishing activity is reliable enough, since we've got high values of reliability coefficients. Reliability for items is lower, than for HEEs, because we should have more items in our experiment. These results are the example of measurement of a single indicator. The similar models can be used for measurement of other research activities indicators. All these results delivered to HEE management can be useful for planning the further research process and improvement of the current situation.

5   Conclusions

In this paper we described assessment models underlying the software component of monitoring and evaluation information system. They are the HEE model (which defines the properties of business processes that have to be measured), the data sources model (which represents our knowledge about information allocation on the web), the scoring model (which supports a procedure of transformation of data from the web into scores), the measurement model (which provides scores processing into values of the desired properties). The described models are validated on the example of university research business process and its properties. We highlighted the problem of measurement model development, since it is the most important and calculation-intensive. We suggested to use Rasch model and substantiated the probabilistic dependency between its variable. According to this model HEE's scores are estimated in dichotomous scale. However, there are other Item Response Theory models, that extend Rasch model. For example, we can explore two- and three-parametrical Birnbaum models, Partial Credit model and others [12].

As one of the future directions of our research we determine investigation of Partial Credit model as a measurement model in the described framework. This model supports polytomous scale of scores. It introduces the grades of successful item accomplishment. We suppose that the HEE measures have to be more differentiable in the case of usage of this model. So our future work is related to additional experiment and further discussion of the results of Item Response Theory models application.

The considered models must be integrated into the monitoring process of HEE quality management system. Therefore we are planning to continue our work in this direction. The case study of this work represents the measurement of a single property of research process, namely, the activity of research results publishing. Further we need to describe all properties of research business process in the university, define the values which should be measured, develop data sources models for each property, and finally, provide the monitoring system with values of the properties. The obtained results can be used as a feedback of management activities and can influence their improvement.

References

1. Barker, C., Pistrang, N., Elliott, R.: Research Methods in Clinical Psychology. An Introduction for Students and Practitioners, 2nd edn. Atrium, Wiley (2002)
2. Okes, D., Westcott, R.T.: The Certified Quality Manager Handbook, 2nd edn. ASQ Quality Press, Milwaukee (2002)
3. Cherednichenko, O., Yangolenko, O.: Towards Higher Education Quality Assessment: Framework for Students Satisfaction Evaluation. In: Proc. of 4th International Conference on Computer Supported Education (CSEDU 2012), vol. 2, pp. 108–112. SciTePress (2012)
4. Cherednichenko, O., Yanholenko, O., Liutenko, I.: Issues of Model-Based Distributed Data Processing: Higher Education Resources Evaluation Case Study. In: Proc. 8th Int. Conf. ICTERI 2012, pp. 147–154. CEUR-WS (2012)
5. Cherednichenko, O., Yanholenko, O., Liutenko, I., Iakovleva, O.: Monitoring and Evaluation Problems in Higher Education: Comprehensive Assessment Framework Development. In: Proc. of the 5-th Int. Conf. on Computer Supported Education, CSEDU 2013, pp. 455–460. SCITEPRESS (2013)
6. Cherednichenko, O., Yanholenko, O., Iakovleva, O.: Web-Based Monitoring and Evaluation: Research Activity Assessment Case Study. In: Proc. in SCIECONF 2013, pp. 455–458. EDIS Publishing Institution of the University of Zilina, Zilina (2013)
7. Cherednichenko, O., Yangolenko, O.: Towards Quality Monitoring and Evaluation Methodology: Higher Education Case-Study. In: Mayr, H.C., Kop, C., Liddle, S., Ginige, A. (eds.) UNISON 2012. LNBIP, vol. 137, pp. 120–127. Springer, Heidelberg (2013)
8. Kusek, J.Z., Rist, R.C.: Ten Steps to a Results-Based Monitoring and Evaluation System: a Handbook for Development Practitioners. The World Bank, Washington, DC (2004)
9. Cherednichenko, O., Yanholenko, O.: Towards Web-Based Monitoring Framework for Performance Measurement in Higher Education. Science and Education a New Dimension: Natural and Technical Science 8, 151–155 (2013)

10. QS World university rankings by subject Computer Science & Information Systems 2014 (2014), http://www.topuniversities.com/university-rankings/university-subject-rankings/2014/computer-science-information-systems
11. Bondarenko, M.F., Shabanov-Kushnarenko, U.P.: Theory of intelligence: a Handbook. SMIT Company, Kharkiv (2006)
12. Steyer, R., Smelser, N.J., Jena, D.: Classical (Psychometric) Test Theory. Elsevier Science (2001)
13. Wright, B., Stone, M.: Measurement Essentials, 2nd edn. Wide Range Inc., Wilmington (1999)
14. Cherednichenko, O., Yanholenko, O., Norbutaev, A.: Web-Based Monitoring: Multiagent Implementation of Data Sources Searching. In: Proc. of the 2nd Global Virtual Conference 2014, GV-CONF 2014 (2014)
15. Top journals in Computer Science, http://academic.research.microsoft.com/RankList?entitytype=4&topDomainID=2&subDomainID=0&last=0&start=1&end=100

Author Index